STP 1013

Effects of Soil Characteristics on Corrosion

Victor Chaker and J. David Palmer, editors

ASTM

1916 Race Street

Philadelphia, PA 19103

Library of Congress Cataloging-in-Publication Data

Effects of soil characteristics on corrosion/Victor Chaker and J.
   David Palmer, editors.

   (STP; 1013)
   "Papers presented at the symposium of the same name held in
Cincinnati, OH on 12 May 1987. The symposium was sponsored by ASTM
Committee G-1 on Corrosion of Metals."—Fwd.
   Includes bibliographies and index.
   "ASTM publication code number (PCN) 04-010130-27"
   ISBN 0-8031-1189-4
   1. Soil corrosion—Congresses. I. Chaker, Victor. II. Palmer,
J. David. III. American Society for Testing and Materials.
Committee G-1 on Corrosion of Metals. IV. Series: ASTM special
technical publication; 1013.
TA418.74.E33   1989   88-38337
620.1′623—dc19   CIP

NOTE

The Society is not responsible, as a body,
for the statements and opinions
advanced in this publication.

Peer Review Policy

   Each paper published in this volume was evaluated by three peer reviewers. The authors
addressed all of the reviewers' comments to the satisfaction of both the technical editor(s) and
the ASTM Committee on Publications.
   The quality of the papers in this publication reflects not only the obvious efforts of the authors
and the technical editor(s), but also the work of these peer reviewers. The ASTM Committee on
Publications acknowledges with appreciation their dedication and contribution of time and effort on behalf of ASTM.

Printed in Ann Arbor, MI
February 1989

Foreword

This publication, *Effects of Soil Characteristics on Corrosion,* contains papers presented at the symposium of the same name held in Cincinnati, OH on 12 May 1987. The symposium was sponsored by ASTM Committee G-1 on Corrosion of Metals. Victor Chaker, The Port Authority of New York & New Jersey, and J. David Palmer, Corrosion Control Engineering, Ltd., presided as symposium chairmen and were coeditors of this publication.

Contents

Introduction

Corrosion of metals in soils is responsible for a large percentage of corrosion worldwide. Several individual characteristics have been used to indicate the corrosivity of soils. However, no documentation describes the synergistic effect of several soil characteristics. This led Subcommittee G1.10 on Corrosion in Soils to create a task group to discover the answer. The task group decided to sponsor an international symposium to find the latest activities in the field of corrosion of metals in soils. The symposium was held 12 May 1987 in Cincinnati. Eleven papers were presented, followed by question and answer sessions.

The symposium revealed specific projects that are being carried on. Several papers expanded the knowledge of one parameter: oxygen concentration cells and their effect on concentric neutral cables. The most promising work was in a paper in which many soil characteristics were correlated using statistical analysis. The technical contributions of each paper are highlighted in the Summary in the back of the book.

More work is needed in the field of corrosion of metals in soils. Such information could be very important in identifying the synergistic effect of all the synergistic parameters, leading to more technically and economically effective methods of corrosion control.

On behalf of ASTM Committee G-1, Subcommittee G1.10, and the Task Group, I wish to express my sincere gratitude to the authors and technical reviewers who made this publication possible.

Victor Chaker

The Port Authority of New York & New Jersey, Jersey City, NJ 07310-1397; symposium co-chairman and editor

John H. Fitzgerald, III[1]

The Future as a Reflection of the Past

REFERENCE: Fitzgerald, J. H., III, **"The Future as a Reflection of the Past,"** *Effects of Soil Characteristics on Corrosion, ASTM STP 1013,* V. Chaker and J. D. Palmer, Eds., American Society for Testing and Materials, Philadelphia, 1989, pp. 1-4.

ABSTRACT: This symposium will deal with the various parameters that affect soil corrosivity and the methods used to measure those parameters. Each soil characteristic has been used at some time in the past to indicate corrosivity, and methods were developed for its measurement. This talk will delve into the background of these parameters and measurements methods and show how past studies affect future understanding. Where did the 10 000 ohm centimeter plateau come from? What was the contribution of the National Bureau of Standards? How did Starkey and Wight affect modern thinking on bacteriological corrosion? What did we learn from the Shepard cane, Columbia rod, and Putman's apparatus? Who was Wenner, anyway?

KEY WORDS: acidity, bacteriological corrosion, history, redox, resistivity, soil corrosion, soil maps, statistical analysis

How It All Started

Alvin Toffler, in his book *Future Shock,* states that scientific knowledge doubles every ten years. When one considers the increase in knowledge since the beginning of the industrial revolution, that statistic becomes somewhat overwhelming. Even if we look back over the relatively short history of corrosion control, we find that about 65 times as much is known today about corrosion as was known in 1930.

And we keep on learning. But we also build upon knowledge gained in years gone by. So, before we begin this symposium, let us pause for a moment and consider the activities and contributions of those who have gone before us. How is their work reflected in what we are doing today?

At the turn of the century, all corrosion was attributed to stray current from rail traction systems—trolley cars and subways [1]. In 1910, Congress authorized the National Bureau of Standards to begin a study of this "stray current electrolysis" that was causing so much damage. In the course of its study, however, the Bureau discovered that corrosion would also occur in soils where no stray current was present. By 1920, NBS had concluded that soil corrosion was equally as serious as corrosion caused by stray current. So the study was expanded in 1922 to determine the causes of soil corrosion; finding that some soils were more corrosive than others, the Bureau went on to determine just what soil parameters were responsible for the corrosion of metals.

What NBS found out is something that we have all come to appreciate in later years, that the corrosivity of a particular soil is based upon the interaction of several parameters—resistivity, dissolved salts, moisture content, pH, presence of bacteria, amount of oxygen, and others. No one parameter can be taken as indicative of the corrosivity of a given soil. The results of the

[1]Vice president, PSG Corrosion Engineering, Inc., The Hinchman Co., 1605 Mutual Building, Detroit, MI 48226.

studies were presented in Circular C450, *Underground Corrosion* [2] in 1945. In it, the Bureau states succinctly that soil corrosion is too complex to permit correlation with any one parameter.

The Bureau's burial tests had established the severity of corrosion to be expected on various metals in various soils. The data were very useful if one were working with a soil similar to one in which tests had been made. But what if one were working with a different soil? It obviously would not be feasible to wait for the outcome of a burial test. It became apparent, therefore, that perhaps one could obtain data on soil parameters and compare those data with NBS test results and thereby come up with an estimate of the soil's corrosivity.

This is where it all began. Today, we are able to measure the pertinent soil parameters and indeed make reasonable estimates about the corrosivity of a particular soil to a buried metallic structure. But who laid the groundwork for the tests we so commonly perform in the field and laboratory today? How is their work reflected in today's terminology and instrumentation?

Resistivity

In 1916, F. Wenner [3] demonstrated that the resistivity of a volume of earth could be measured by inserting four pins into the earth in a straight line, passing a current between the two outer two pins and measuring the resultant resistance between the inner two pins. Known as the "Wenner four pin method," this technique is one of the most common methods of measuring soil resistivity today. Early instruments used with this setup were the McCollum earth current meter and the hand-cranked Megger. Modern, solid state instruments may be faster and more accurate, but they perform about the same function as the old timers did!

Another early instrument that helped in obtaining soil resistivity, particularly in relatively confined areas, was the Shepard cane. Developed by E. R. Shepard [4] about 1930, the instrument consisted of two rods about three feet long, each tipped with a iron electrode. The rods were inserted in holes in the ground about 8 or 10 in. apart, and a current from a 3-V battery was passed between them. The instrument contained an ammeter that was calibrated in ohm centimeters. From the Shepard cane has come several probe-type instruments for the measurement of soil resistivity.

Soil resistivity is relatively easy to measure and, being an electrical quantity and thus related to corrosion current flow through Ohm's law, is probably the parameter most often looked upon as indicative of a soil's corrosivity. Even today we see codes that state that corrosion protection is not required in soils of resistivity above 10 000 ohm centimeter or some similar figure. Where did that plateau come from? It goes back to 1940s; it had been found that, in soils of resistivity above 10 000 ohm centimeter, the rate of corrosion on pipelines was generally slow enough that it was less expensive to repair leaks when they occurred than to provide corrosion protection. That may have been all right in the political atmosphere of those days, but it is no longer valid. Bacteria, dissimilar metals, or oxygen concentration cells, for example, may create severe corrosion in high-resistivity soil; we are concerned today about safety, pollution, and economics and can no longer rely on simple measurement as a basis for determining the need for corrosion control.

Acidity

Often attempts are made to relate corrosivity to the pH of the soil; people worry about "acid" soil. The question goes back to 1924 when J. W. Shipley, I. R. McHaffie [5], and H. D. Holler [6] observed that there appeared to be a relationship between soil acidity and the rate of corrosion of iron. It remained for I. A. Denison and R. B. Hobbs [7] in 1934, however, to determine that corrosion was related to the total acidity of the soil rather than just the pH. In 1935 I. A. Denison and S. P. Ewing [8] discovered a relationship between total acidity and soil resistivity in tests in northern Ohio. The complexity of soil corrosion was beginning to be appreciated as experimenters began to realize that no one soil parameter could tell the whole story.

Bacteriological Corrosion

Bacteriological corrosion, a serious concern in many parts of the country today, came under investigation as early as 1923 when R. Stumper [9] reported on the corrosion of iron in the presence of sulfur. Early concern centered around the corrosion of pipes having joints caulked with sulfur-bearing compounds. The matter came under further investigation along about 1940. R. L. Starkey and K. L. Wight [10] came to the conclusion that oxidation-reduction (redox) potential was the most reliable indicator of bacteriological action. They developed what was later to become known as the redox probe. Consisting of two units, one containing a platinum and calomel electrode and the other a calomel and a glass electrode (to measure pH), the instrument was later refined into a single probe for improved mobility and field use.

Interest in bacteriological corrosion began to grow in the succeeding two decades as a possible cause of corrosion failures in high-resistivity or other soils that would otherwise be thought of as only mildly corrosive. Extensive research was undertaken by J. O. Harris [11], who studied the action of bacteria under both aerobic and anaerobic conditions. He also did some rather controversial work on the effect of bacteria on coatings. Harris' work, combined with that of earlier researchers, helped pave the way to a modern understanding of the phenomenon of bacteriological corrosion.

Interrelation of Soil Parameters

All of this so far has dealt with the development of understanding of the contribution to corrosion of various soil parameters. There remains the question of how the parameters react together to produce corrosion. The question was pondered as early as 1930 when B. B. Legg [12] developed the Columbia rod. This was a probe-type instrument having a steel and a copper electrode at its tip, connected through a milliammeter. It was an attempt to correlate soil resistivity, potential differences pH, and polarization. The data obtained gave an indication of corrosivity when compared against observed pipeline conditions in soils of known resistivity. J. F. Putman [13] in 1917 developed a device to measure soil resistivity and modified it in 1935 in an attempt to combine pH and resistivity as a measure of what he called the "potential corrosivity" of the soil. The complexity of soil corrosion was indeed beginning to be realized.

Today, one method of evaluating soil corrosivity is through statistical analysis of soil characteristics and pit depths on pipelines. Here we look back on a heritage developed in the hydroelectric industry over what would be the maximum flood level or water flow over the expected lifetime of a generating plant or dam. Engineers spoke of the "hundred year flood" and added arbitrary safety factors such as perhaps double the size of the largest flood that had occurred during the past 50 years. In the mid-1950s, E. J. Gumbel [14] applied extreme value statistics to flood evaluation and was able to predict the probability of a major flood occurring every so many years. G. G. Eldridge [15] applied this concept to pitting of oil well tubing in the late 1950s and his work has been expanded into the procedures used today to evaluate pipe condition. Also in the late 1950s, G. N. Scott [16] used extreme value statistics to evaluate soil resistivities, furthering our understanding of the chances of encountering low-resistivity soil at a given site.

Soil Maps

We would be remiss if we didn't mention soil maps. From time to time various maps have appeared purporting to show areas of "corrosive" and "noncorrosive" soils. It all began in 1899 when the U.S. Department of Agriculture began mapping the soils of the United States. While not addressing corrosivity, the U.S.D.A. reports do cover aeration, drainage, and other soil characteristics useful to the corrosion engineer. C. F. Marbut [17] in 1935 classified the soils of the United States into eight great groups and identified many subgroups.

These maps can be helpful in preliminary planning, but we still must recognize that soils of similar types in different areas may not exhibit the same corrosion characteristics. The corrosion engineer still needs to get out in the field and evaluate what is to be expected at a given site or along a specific right of way. There may also be a considerable amount of corrosion data available to assist in the investigation which, when combined with newly acquired field data, will give a good indication of the corrosivity of the soil.

Epilogue

Much has been accomplished in the past; much remains to be learned in the future. Ten years from now we will know twice as much as we do now. Perhaps we will have instrumentation that can combine all the soil parameters and give us a good indication of the effect of soil on various metals. But let us not forget that many investigators have postulated equations and formulae over the past years in attempts to do the same thing!

Thomas a'Kempis, a German monk of the fifteenth century, said, "Today I pray for the wisdom to build a better tomorrow on the mistakes and experiences of yesterday." Let us apply that same philosophy to corrosion engineering, and as we go forward in our understanding of soil corrosion let us remember and apply the contributions of those who have labored in the past to bring us to where we are today.

References

[1] Romanoff, M., *Underground Corrosion,* NBS Circular 579, National Bureau of Standards, Washington, DC, 1957.

[2] *Underground Corrosion,* NBS Circular C450, National Bureau of Standards, Washington, DC, 1945.

[3] Wenner, F., "A Method of Measuring With Resistivity," *Bureau of Standards Scientific Papers,* Vol. 12, S258, 1916.

[4] Shepard, E. R., "Pipeline Currents and Soil Corrosivity as Indicators of Local Corrosive Areas," *Bureau of Standards Journal of Research,* Vol. 6, RP 298, 1931.

[5] Shipley, J. W. and McHaffie, I. R., "The Relation of Hydrogen Ion Concentration to the Corrosion of Iron," *Chemistry and Metallurgy,* Vol. 8, No. 121, Canada, 1924.

[6] Holler, H. D., "Corrosiveness of Soils with Respect to Iron and Steel," *Industrial Engineering Chemistry,* Vol. 21, No. 750, 1929.

[7] Denison, I. A. and Hobbs, R. B., "Corrosion of Ferrous Metals in Acid Soils," *Journal of Research,* NBS 13, No. 125, RP696, 1934.

[8] Denison, I. A. and Ewing, S. P., "Corrosiveness of Certain Ohio Soils," *Soil Science,* Vol. 40, No. 287, 1935.

[9] Stumper, R., "La Corrosion du Fer en Presence du Sulfure du Fer," *Compt. rend. Acad. Aci.* (Paris), Vol. 176, No. 1316, 1923.

[10] Starkey, R. L. and Wight, K. L., "Anaerobic Corrosion of Iron in Soils," final report of the American Gas Association Research Fellowship, Monograph AGA, 1945.

[11] Harris, J. O., "Microbiological Studies Reveal Significant Factors in Oil and Gas Pipeline Backfilled Ditches," Technical Bulletin 135, Kansas Agricultural Experiment Station, Kansas State University, Manhattan, KS, 1963.

[12] Legg, B. B., "Early Steps in the Development of the Columbia Soil Rod," *Gas Age Record,* Vol. 67, No. 111, 1931.

[13] Putman, J. F., "Electrolysis," *Sibley Journal of Engineering,* Vol. 31, No. 88, 1917.

[14] Gumbel, E. J., "Statistical Theory of Extreme Values and Some Practical Applications," NBS Applied Mathematics Series 33, U.S. Dept. of Commerce, Washington, DC, 1954.

[15] Eldridge, G. G., "Analysis of Corrosion Pitting by Extreme Value Statistics and Its Application to Oil Well Tubing Caliper Surveys," *Corrosion,* Vol. 13, No. 1, 51t, 1957.

[16] Scott, G. N., "The Distribution of Soil Conductivities and Some Consequences," *Corrosion,* Vol. 14, No. 8, 396t, 1958.

[17] Marbut, C. F., "Atlas of American Agriculture, Part III: Soils of the U.S.," U.S. Government Printing Office, Washington, DC, 1935.

J. D. Palmer[1]

Environmental Characteristics Controlling the Soil Corrosion of Ferrous Piping

---

REFERENCE: Palmer, J. D., **"Environmental Characteristics Controlling the Soil Corrosion of Ferrous Piping,"** *Effects of Soil Characteristics on Corrosion, ASTM STP 1013,* V. Chaker and J. D. Palmer, Eds., American Society for Testing and Materials, Philadelphia, 1989, pp. 5–17.

ABSTRACT: The characteristics of soils controlling the external corrosion of ferrous piping materials are examined with particular reference to an American Water Works Association (AWWA) rating formula. The relationship and reliability of the following characteristics as corrosion predictors are examined: resistivity, pH, redox potential sulfides, moisture, and chlorides. Metallurgical test results illustrating the mechanism of attack on cast iron and ductile iron are presented and the apparently different performance of the two materials is reviewed. Mitigative and preventive measures are mentioned and the organization of a corrosion control program outlined.

KEY WORDS: underground corrosion, piping materials, parameters, resistivity, pH, redox, moisture, surveys, mitigation

Over the last century the material used for water (and gas) mains has changed from wood to pit-cast iron to centrifugally cast iron to ductile iron to plastic with occasional introduction of wrought iron, plain and galvanized steel, coated and cathodically protected steel, asbestos cement, and reinforced concrete. Customer service piping has been made from all of the above materials as well as from copper. Current interest is focusing on the performance of ductile iron, which largely replaced cast iron beginning in the 1960s. Ductile iron failure rates of 0.5 leaks/km/year are common, with Winnipeg recording even higher rates beginning in the 1950s [1].

Material Performance

Wood must be considered a primitive material with the early butt-joint and strapped wood stave piping limited in pressure and subject to rot and attack by fungi.

Cast iron (pit-cast and centrifugally cast) has commonly given service life in the 100-year range, and post-World War II studies showed percentages of installed pipe still in service in the 90% range, increasing with increased pipe size [2]. Studies conducted in the early 1960s did not show the exponential increase in leak (break) rate commonly associated with corrosive attack, but later studies have shown this effect, suggesting that the corrosivity of the soils has increased since the 1960s.

The interpretation of cast iron failure data is difficult because most failures are described as "breaks," whether due to purely mechanical effects or partially due to the weakening effect of corrosion. Usually the only leaks attributed to corrosion are those where there is an obvious blowout of the graphitized part of the pipe wall without an accompanying mechanical failure.

[1]President, Corrosion Control Engineering, Ltd., London, Ontario, Canada.

The later failure mechanism is more commonly found in ductile iron piping. As subsequently noted, specific inspection procedures are necessary for the identification of iron corrosion.

Increasing failure rates coincided with the tremendous increase in the use of road deicing salts in North America, most immediately evident on automobiles beginning with the 1955 model year. The effects were first evident in areas where deicing salt use was common, where the natural salt content of the soil was high, or where the local soil had natural corrosive and mechanical characteristics conducive to early failures. Winnipeg falls into the later classification, and the poor performance of cast iron in Winnipeg was well documented in the 1950s [3].

Asbestos cement was recognized as an alternative in the 1950s, but mechanical considerations resulted in rather limited use and it is subject to destructive leaching in some soils and waters.

Ductile iron was inserted into this situation in the 1960s and was widely used as a replacement for cast iron on the understanding that its corrosion resistance was "equal or superior to grey cast iron" [4]. Unexpectedly, early failures were first identified in the Metropolitan Toronto area in the 1970s, apparently due to the acceleration effects of stray d-c currents produced by the area transit systems. Area pipeline people had been active in combatting stray d-c effects for some years and were then joined by the waterworks people—only recently achieving appreciable mitigation. During this period there was considerable debate concerning the relative performance of cast and ductile iron. Currently, many municipalities have instituted extensive programs for the identification of high-risk piping and mitigation by a combination of cathodic protection supplemented by bonding where stray dc is a contributing factor. Most new transit systems use insulated rail systems which greatly alleviate the stray current situation.

Although the hydrocarbon-transportation industry quickly introduced cathodically protected coated steel pipe beginning in the 1950s because of its improved pressure rating and relative economy of installation, only limited installations were made by waterworks people. In general, these installations were of limited success because of the failure to maintain the electrical isolation necessary for effective cathodic protection.

Plastic piping took over an appreciable percentage of the market in the 1970s in spite of the insufficiency of long-term service data. Isolated instances of early failures have begun to appear and a judgement on suitability cannot yet be made. Reinforced concrete pipe has been used for high-pressure transmission systems, and failures have been experienced in high-salinity soils. In view of the disastrous performance of reinforced concrete in highway and parking structures, the performance of reinforced concrete in deicing salt-contaminated soil must be considered suspect.

Corrosion Morphology—Ferrous Materials

The corrosion of mild steel produces no particularly significant behavior other than the usual lowering of corrosion rate with time as the corrosion products introduce additional resistance in the corrosion cell electrical circuit. Cast iron has been characterized by the significant pressure-retaining ability of the corrosion product, with perforated pipe retaining appreciable pressure capacity [4]. This characteristic has also been attributed to ductile iron by some sources [5].

The relationship of the cathodic (noncorroding) graphite to the anodic (corroding) iron in cast and ductile iron pipe has long been of interest, and the size and shape of the graphite particles in relation to corrosion resistance have been examined—without firm conclusions being reached.

An extended series of physical inspections and metallurgical tests of Metro Toronto gas mains examined this aspect [2]. The in-the-ditch inspections and laboratory examination of pipe samples confirmed that the cast iron corrosion products were hard and dense—often resembling the natural pipe surface and requiring chipping or sandblasting for identification. Our examinations of ductile iron pipe have revealed similar characteristics.

Older cast iron pipe is typified by considerable variation in graphite flake size, covering the complete range of ASTM size classification (Fig. 1) [ASTM Method for Evaluating the Microstructure of Graphite in Iron Castings (A 247-67)]. Reflecting improved technology and individual foundry practice, newer centrifugally cast pipe tends to have smaller and more uniform graphite flakes (Fig. 2). Ductile iron appears to be produced with more uniform procedures, producing uniform graphite nodules (Fig. 3).

Metallurgical tests tend to confirm that the corrosion of cast iron, a form of "dealloying corrosion," is nucleated by the graphite-iron galvanic cell and suggested that the graphite/corrosion product deposits pressure retaining ability is influenced by the characteristics of the matrix established by the graphite flakes [2]. There was also a suggestion that the deposit formed becomes increasingly dense as soil resistivity increased, perhaps due to the reduced ability of the iron ion to migrate away from the corrosion site. Neither of these effects were firmly established and are not covered in the published literature.

Where cast and ductile iron corrosion is accelerated by stray d-c currents, it would appear that the corrosion products are transported away from the corrosion site, and the classical inplace graphitic corrosion product is not evident to the same degree as with normal corrosion.

Ductile iron has been reported to have better corrosion resistance than cast iron, but the exposures were generally conducted over relatively short periods in very low (<500 ohm-cm) resistivity soil [6]. A review conducted by Canada's National Research Council concluded that the corrosion rate of all the ferrous materials by soils is essentially equal [7]. The high early failure rates for ductile iron can then be explained by the lower thickness for the same pressure rating as shown in Fig. 4. Assuming the normal exponential reduction in penetration rate, if cast iron perforates in 50 years, the equivalent pressure rating ductile iron would perforate in 18 years. The even thinner schedule 40 mild steel would perforate in 15 years.

Mitigative Action

To date, mitigative action for ductile iron has consisted mainly of logging leaks and installing sacrificial cathodic protection anodes at each leak. A preventive program based on historical leak data can be established and anodes installed on a planned basis in augered holes using cleaning and attachment techniques which permit the work to be done in a small hole, working from grade. Such programs have been described at many ASTM, AWWA, and NACE (National Association of Corrosion Engineers) meetings and a number of proprietary approaches have been developed.

At repair costs which may exceed $2,000 Canadian/leak, there is considerable economic advantage in the early initiation of preventive programs involving a combination of soil mapping, pipe size-age plotting, and the planned installation of sacrificial anodes based on risk of a leak.

Soil Characteristics

A large number of variables has been given consideration in AWWA Standard C 105-72 (Table 1), proposed by the Cast (Ductile) Iron Pipe Research Association (CIPRA) and widely used by waterworks departments in selecting pipe materials and protective measures. Each of these parameters is reviewed with respect to its reliability and relevance as a corrosion indicator.

Resistivity

Resistivity, the reciprocal of conductivity, indicates the ability of an environment to carry corrosion currents. Most pipe corrosion cells tend to be local, and the resistance of the pipe is very much lower than the resistance of the soil path. Resistivity is a function of the soil moisture and the concentration of current-carrying soluble ions. Measured in ohm-cms, resistivity can

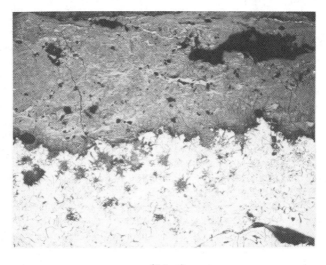

(56 x)

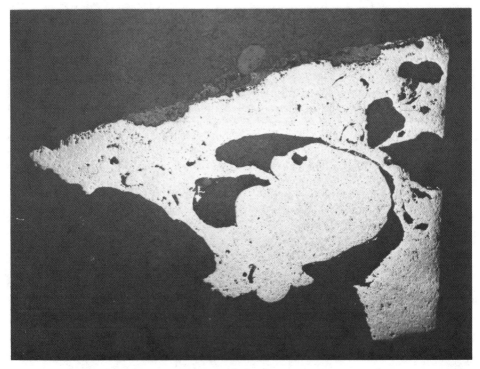

(7.5 x)

FIG. 1—*Fifty-one-year-old CI pipe—resistivity >10 000 ohm-cm—showing casting defects and corrosion nucleation around fine graphite flakes.*

(56 x)

(7.5 x)

FIG. 2—*Fifty-five-year-old cast iron pipe—resistivity 500 ohm-cm—showing corrosion nucleation around coarse graphite flakes.*

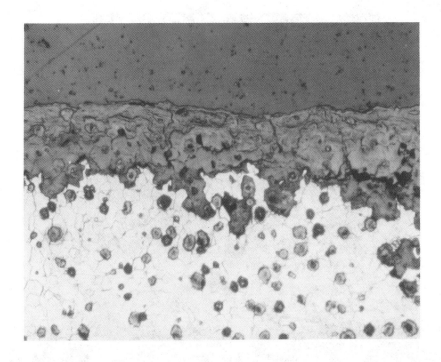

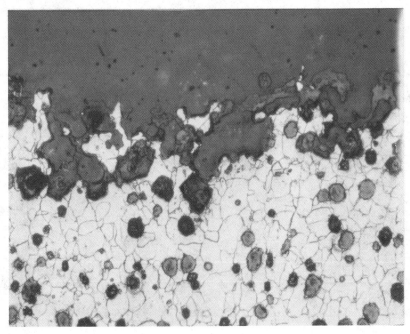

FIG. 3—*Five-year-old ductile iron pipe—230 ohm-cm soil showing corrosion nucleation around graphite nodules (X250) of ductile iron.*

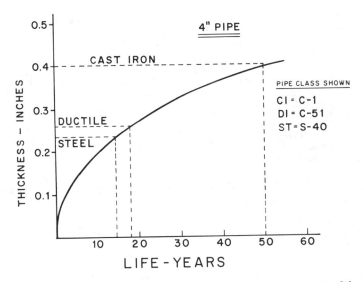

FIG. 4—*Typical exponential corrosion rate relationship for ferrous pipe* [1].

vary from 30 ohm-cm in seawater to in excess of 100 000 ohm-cm in dry sand or gravel. The AWWA formula considers less than 700 ohm-cm to be severely corrosive, while the steel pipeline industry considers anything less than 1000 ohm-cm to be "very severely corrosive" (Table 2) [8]. This difference may reflect the tendency for pitting rates to be higher at the defects in coated pipe than on bare pipe as a result of the decreased anode/cathode area ratio. Using the AWWA formula, if the pipe is to be wet, the rating points are the same. The overwhelming majority of field studies show resistivity to be the major controlling parameter except for areas with severe microbiological activity [9].

Resistivity may be measured at grade, but, as pointed out in ASTM Method for Field Measurement of Soil Resistivity Using the Wenner Four-Electrode Method (G 57-78), contamination effects may be missed and the most relevant resistivity measurement is obtained from a sample taken from the pipe ditch (Fig. 5). Where a volume of at-grade and in-the-ditch data has accumulated, at-grade measurements from undeveloped areas can be adjusted to reflect the effect of future contamination. Our experience has shown that single-probe measurements are totally unreliable. Only the Wenner four-pin method can be recommended. Statistical analytical techniques best display the significance of the data obtained [8]. In deicing salt areas, chloride contamination appears to be the main factor in increased soil corrosivity with levels in excess of 0.01% considered indicative of accelerated corrosion. As well as reducing resistivity, the chloride ion tends to break down otherwise protective surface deposits and can result in the cracking of stressed stainless steel hardware such as leak clamps.

pH

Acidity, indicated by the pH value, is given considerable weight in the CIPRA formula, but only when lower than 4 or higher than 8.5. In the pH range of 4 to 8.5, iron can be immune (not corroding), passive (corroding very slowly), or corroding actively, depending on its potential, as shown by a simplified Pourbaix diagram (Fig. 6). Due to the leaching effect of rainfall and the presence of acid rain, most eastern soils tend to be somewhat acidic, but rarely with a pH lower

TABLE 1—*Soil-Test Evaluation AWWA Rating—Standard C 105-72.*

Soil Characteristics	Points

RESISTIVITY—OHM-CM
(based on single probe at pipe depth or water-saturated Miller soil box)

<700	10
700 to 1000	8
1000 to 2000	5
1200 to 1500	2
1500 to 2000	1
>2000	0

pH

0 to 2	5
2 to 4	3
4 to 6.5	0
6.5 to 7.5	0
7.5 to 8.5	0
>8.5	3

REDOX POTENTIAL

>+100 mV	0
+50 to +100 mV	3.5
0 to + 50 mV	4
Negative	5

SULFIDES

Positive	3.5
Trace	2
Negative	0

MOISTURE

Poor drainage, continuously wet	2
Fair drainage, generally moist	1
Good drainage, generally dry	0

TABLE 2—*Steel pipe corrosion classification.*

0 to 1 000	ohm-cm very severely corrosive
1 001 to 2 000	ohm-cm severely corrosive
2 001 to 5 000	ohm-cm moderately corrosive
5 001 to 10 000	ohm-cm mildly corrosive
>10 001	ohm-cm very mildly corrosive

than 4 unless there is severe industrial contamination. A neutral pH [7] is the most favorable level for sulfate-reducing bacteria (see section entitled Redox Potential).

The natural potential of cast iron tends to fall in the 0.5 to 0.6 V negative range with reference to a Cu/CuSo$_4$ reference cell. This is somewhat less negative than the usual coated steel potential, reflecting the surface presence of the more noble graphite. It is not our intent to discuss cathodic protection criteria in this presentation, but it is evident from the Pourbaix diagram that immunity may be conferred by a voltage shift in the negative direction.

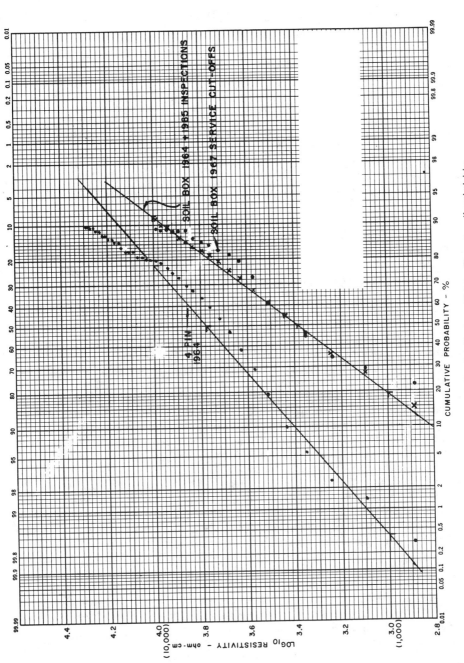

FIG. 5—Cumulative probability distribution of Metro Toronto soil resistivities.

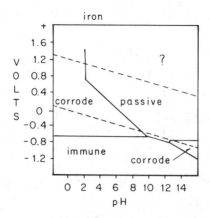

FIG. 6—*Simplified Pourbaix diagram for iron.*

pH measurements may be useful in identifying unusual soil conditions but in most cases are only significant in distinguishing between otherwise similar soils. ASTM Test Method for pH of Soil for Use in Corrosion Testing (G 51-77) indicates that soil pH should be measured in situ or immediately after a sample is removed from the field.

Redox Potential

This parameter attempts to distinguish between aerobic soils and anaerobic soils that could support sulfate-reducing bacterial activity. These bacteria, typified by the Desulphovibro genus, reduce sulfates to sulfides and may activate cathodic areas by consuming hydrogen or may produce corrosive products. Microbiological activity—commonly reported in Europe—is rarely identified as the cause of corrosion in North America. The test, to be valid, must be conducted in situ or immediately after soil exposure and requires careful electrode preparation and calibration. These conditions are rarely met and most reported Redox potentials are quite positive (> 100 mV), indicative of aeration of the soil sample during removal and/or transportation to a laboratory. Microorganisms cultured by aerobic conditions may similarly stimulate anodic areas by oxygen absorption and acid production.

Kuhlman and others have attempted to correlate Redox potential with corrosion rate but their efforts have generally been unsuccessful. Kuhlman's data are much more reliable when plotted against resistivity, as shown in Fig. 7 [*10*]. As for pH, this parameter is normally only useful in distinguishing between otherwise identical soils and then only when properly measured. Other test methods are more reliable in identifying bacterial activity.

Sulfides

Most soils will show at least a trace of sulfides and/or sulfates, and this only may be significant in conjunction with the relevant redox potential (< +100 mV). Sulfate levels are of more significance where concrete structures are concerned.

Moisture

The pipe ditch acts as a subsurface drain, collecting surface moisture and contaminants, and, from a practical standpoint, most pipes must be considered to be continuously wet—at

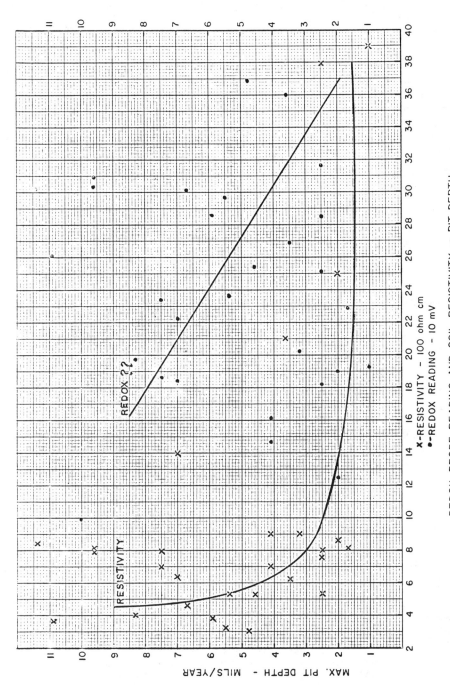

REDOX PROBE READING AND SOIL RESISTIVITY vs PIT DEPTH

x-RESISTIVITY - 100 ohm cm
•-REDOX READING - 10 mV

FIG. 7—*Comparison of correlation of redox reading and resistivity with pit depth.*

least on the bottom. Present procedures for moisture content determination require the oven-drying of a soil sample.

Discussion

The AWWA formula requires protective action for cast and ductile iron if the additive total formula points equal ten. If we deduct the points allowed for irrelevant parameters as shown*, only considering pH and resistivity, and presume pH values in the 4 to 8.5 range, we are left with a requirement for protection at three points, or at a resistivity less than 1400 ohm-cm. Values below this level are regularly found in salt-contaminated clay soils, and it follows that cast and ductile iron often require supplemental protection or that alternative more-resistant materials should be used. The coating and cathodic protection of ductile iron is standard practice in some jurisdictions [1].

*Points Deducted

Redox	3
Sulfides	2
Moisture	2
Total	7

Conclusions

The corrosion of buried ferrous pipe is a problem approaching catastrophic proportions in Canada where mitigative action has not been taken either on installation or postinstallation. Of the variables included in the AWWA formula, only resistivity appears to be generally relevant. The other factors may be pertinent where difference in corrosion rate are experienced with otherwise similar conditions.

Resistivity mapping combined with pipe type/age plotting appears to be the most reliable approach to planning mitigative programs, with the other parameters included in the AWWA formula being given lesser consideration. Chloride determination would be a useful test for soil samples taken from the pipe ditch, while stray current evaluation would be an informative factor in areas where dc-powered traction systems are known to exist. In the latter case, this information can often be obtained from the local gas, electric, and telephone organizations as they are most immediately affected by stray current activity. In areas where appreciable stray d-c currents are present, an Electrolysis Coordinating Committee usually operates and mitigative action is best achieved by cooperating with these groups.

Acknowledgments

The permission of the Consumers Gas Co. to use information about their system is appreciated. Brian Doherty's efforts in providing a ductile iron sample from the Scarborough system are appreciated.

References

[1] Palmer, J. D., "Soil Corrosion in Canada—Current Status," Department of Energy, Mines and Resources, Canada, CANMET Project No. 9-9116, 27 February 1981.
[2] Palmer, J. D., "The Life of Buried Pipe," Ontario Research Foundation, 18 February 1966.
[3] Baracus, A., Hurst, W. D., and Legget, R. F., "Effect of Physical Environment on Cast Iron Pipe," Journal AWWA, December 1955.
[4] Ductile Iron Pipe Catalogue, Canron Ltd. Pipe Division.

[5] Fuller, A. G., "Soil Corrosion Resistance of Grey and Cutile Iron Pipe—A Review of Available Information," British Cast Iron Research Association, Report 1073, May 1972.
[6] "Ductile Iron Pipe," soil corrosion test report, Cast Iron Pipe Research Association, 1964.
[7] Scott, J. D., "Mechanism and Evaluation of Corrosion in Soils, A Literature Review," National Research Council, Division of Building Research, Paper No. 100, June 1960.
[8] Palmer, J. D., "Soil Resistivity—Measurement and Analysis," *Materials Performance,* Vol. 13, No. 1, January 1974.
[9] Williams J., *Bibliography on Underground Corrosion Materials Performance,* Vol. 21, Nos. 1, 2, 3, 1982.

Paul A. Burda[1]

Differential Aeration Effect on Corrosion of Copper Concentric Neutral Wires in the Soil

REFERENCE: Burda, P. A., **"Differential Aeration Effect on Corrosion of Copper Concentric Neutral Wires in the Soil,"** *Effects of Soil Characteristics on Corrosion, ASTM STP 1013,* V. Chaker and J. D. Palmer, Eds., American Society for Testing and Materials, Philadelphia, 1989, pp. 18–35.

ABSTRACT: The objective of this research was to identify the dominant effect of differential aeration mechanisms on corrosion of copper concentric neutrals (CN). Laboratory data indicate that the corrosion of CN wires is accelerated by differential aeration/concentration effects when groundwater with mud and debris fills the conduit. The differential aeration was dependent on diffusion of air, nature of environment, anode to cathode area ratios, and pH.

KEY WORDS: corrosion, anodic and cathodic polarization, differential aeration, diffusion, de-aerated and aerated half cells, underground distribution (UD) cables, concentric neutral (CN) wires, imported backfill[2]

Extensive localized corrosion cells of CN wires are causing failures of underground distribution (UD) cables in a few years. Figure 1 shows a typical failure of CN copper wires found on unenergized underground distribution (UD) cables in conduits, exposed three years in soil. The principle of differential aeration effect—that is, the localization of corrosion at the less aerated zones—is considered by some investigators [1–3] to be one of the most probable mechanisms controlling this deterioration of CN wires in soils.

The pH at the metal interphases, the nature of the environment, the ohmic resistance effect, and the anodic reaction are considered dominant factors of the differential aeration mechanism(s) [4]. Subsequent work by Evans [5,6] and Herzog [7] concluded that diffusion of oxygen controls the output of the differential aeration cell and that the ohmic resistance in the electrolyte path is usually a secondary factor. The greater the oxygen concentration in the cathodic region, the greater the corrosion cell current. Anodic polarization, which may also occur at high corrosion rates, can be neglected in an ordinary case [7].

The objective of this work was to identify conditions where differential aeration mechanisms have a dominant effect on the corrosion of copper CN wires exposed in soil.

Experimental

The experimental arrangement permitted measurement of the corrosion current between local action and cell action while the probe was acting as a cell anode or a cell cathode in the

---

[1]Corrosion engineer, Department of Engineering Research, Pacific Gas & Electric Co., San Ramon, CA 94583.
[2]Granular sandy material commonly used for backfilling of areas around the buried structures before the native soil is introduced.

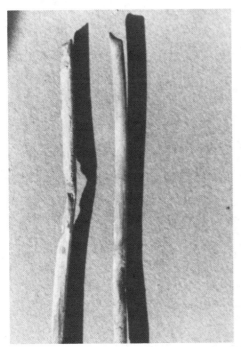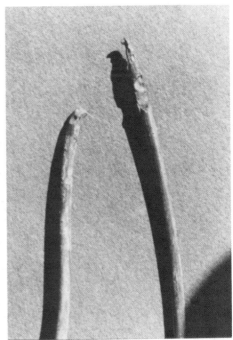

FIG. 1—*Localized corrosion and failures of CN wires on unenergized underground cables exposed three years in soil.*

concentration cell. The model of experimental cell arrangements for the differential aeration study is shown in Fig. 2. Anodic and cathodic areas were separated by porous membranes. Air and nitrogen were purged through one compartment during the exposure. Solutions or moist imported backfill (aerated half cell) formed the cathodic zone, while the moist soil (deaerated half cell) formed the anodic zone. The test probes were prepared from No. 14 AWG bare copper CN wire and were installed in each compartment as shown in Figs. 2 and 3. Both half cells were short-circuited through a small resistor. The resistance value of this shunt was determined from the electrical current balance relationship of prepared cells:

$$\frac{E \text{ resistor}}{R \text{ resistor}} = \frac{E \text{ cell}}{R \text{ cell} + R \text{ resistor}}$$

$$R \text{ cell} = \frac{\text{Resistor} (E \text{ cell} - E \text{ resistor})}{E \text{ resistor}}$$

where

E resistor = voltage across the resistor (volts),
$\ \ \ \ \ E$ cell = open cell potential (volts),
$\ \ \ \ \ R$ cell = inner resistance of the cell (ohms), and
R resistor = shunt resistance (ohms).

The R cells were calculated first, using E cell and galvanic cell current measurements. Then

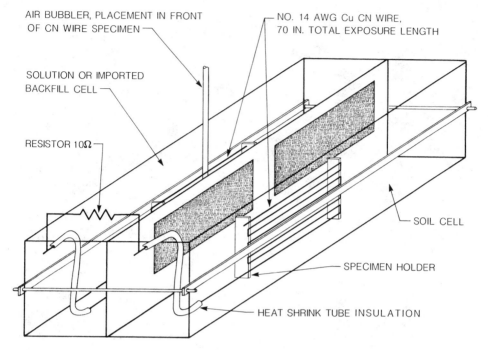

FIG. 2—*Differential aeration measurement on CN wires (cathode to anode ratio 1 : 1).*

the resistance of the shunt (resistor) was chosen to be equal to 10 Ω, approximately one hundredth of R cells. The corrosion current caused by differential aeration was calculated from the potentials measured across the resistor of known resistance using a microvolt meter. Ohm's law was applied for this calculation. Corrosion rates were evaluated using a modified formula of Faraday's law.

$$MPY = \frac{I \times A}{\text{Area} \times \text{Density} \times n} \times 1289.3$$

where

$\qquad I =$ corrosion current (ampere),
$\qquad A =$ atomic weight of exposed material (g/mole),
$\quad \text{Area} =$ exposed material (dm²),
$\text{Density} =$ corroded material (g/cm³),
$\qquad n =$ equivalent per mole, and
$\quad MPY =$ mil per year (1 MPY = 0.001 in. per year = 0.0254 mm per year).

The corrosion potentials were measured against standard calomel electrodes (SCE).

An effect of differential aeration on half cells of the couples was also evaluated by comparing the weight losses on uncoupled and on coupled wires. The surface preparation and cleaning of the probe wires for weight loss corrosion rate evaluation followed procedures given in ASTM Recommended Practice for Preparing, Cleaning, and Evaluating Corrosion Test Specimens (G 1-81).

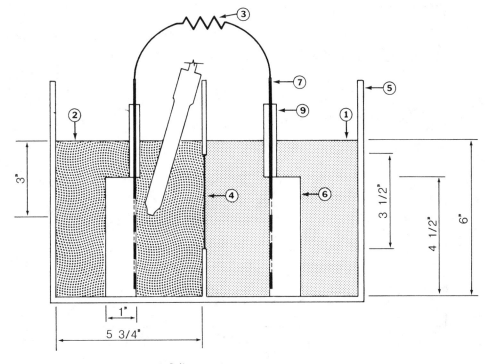

1 Soil
2 Solution or imported backfill
3 Resistor
4 Nuclepore membrane
5 Plexiglass container
6 Specimen holder with CN copper wire
7 Copper wire
8 Saturated calomel electrode (SCE)
9 Heat shrink tubing insulation

FIG. 3—*Differential aeration cell for test of copper CN wire corrosion (front view).*

Results and Discussion

Differential Aeration Effect on Copper CN Corrosion Rate

Table 1 and Figs. 4 through 16 show how our model of differential aeration influenced the corrosion of CN copper wires in soil due to the transfer of electrons from more negative (soil) to more positive (solutions) electrodes of the connected couple, resulting in an increase in corrosion rate of the copper wire in the soil (anode).

The potential of the aerated zone increases and its corrosion decreases owing to passivation. Therefore, when more air is being injected at the aerated cathodic (solution) side, the nonaerated (anodic) corrosion rate increases due to the increase of cathodic passivation (Fig. 4). General corrosion reactions describing this process are: $O_2 + 2H_2O \rightarrow 4OH^-$ (cathodic), $Cu \rightarrow Cu^{++} + 2e^-$ (anodic). Figure 5 shows how the corrosion rate proceeded with time in the presence of chlorides and sulfates.

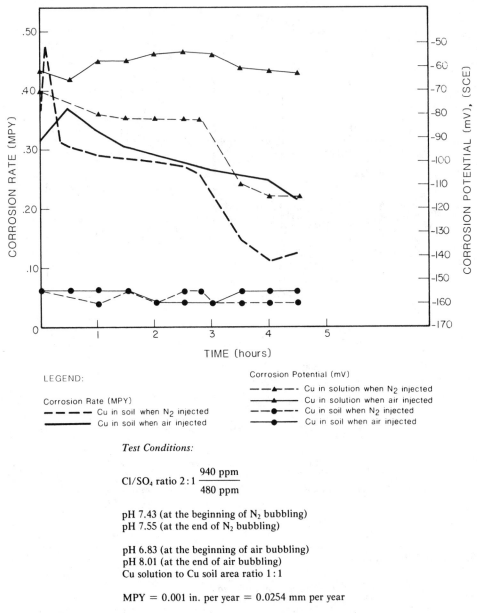

LEGEND:

Corrosion Rate (MPY)
------- Cu in soil when N_2 injected
——— Cu in soil when air injected

Corrosion Potential (mV)
--▲--- Cu in solution when N_2 injected
——▲——— Cu in solution when air injected
--●--- Cu in soil when N_2 injected
——●——— Cu in soil when air injected

Test Conditions:

$$Cl/SO_4 \text{ ratio } 2:1 \quad \frac{940 \text{ ppm}}{480 \text{ ppm}}$$

pH 7.43 (at the beginning of N_2 bubbling)
pH 7.55 (at the end of N_2 bubbling)

pH 6.83 (at the beginning of air bubbling)
pH 8.01 (at the end of air bubbling)
Cu solution to Cu soil area ratio 1:1

MPY = 0.001 in. per year = 0.0254 mm per year

FIG. 4—*Anodic corrosion rate of CN copper wire controlled by differential aeration mechanism in soil and chloride/sulfate solution.*

It was found (Table 1) that the differential aeration effect increased copper corrosion approximately 20 times in the soil. Our findings also revealed (Fig. 6, Table 1) that differential mechanism controlled copper corrosion when CN copper wire was exposed to the backfill and soil. Due to the differential aeration effect, the corrosion in the soil increased approximately ten times. This finding indicates that coarse backfill and clayish soil could be a favorable combination for copper corrosion in soil.

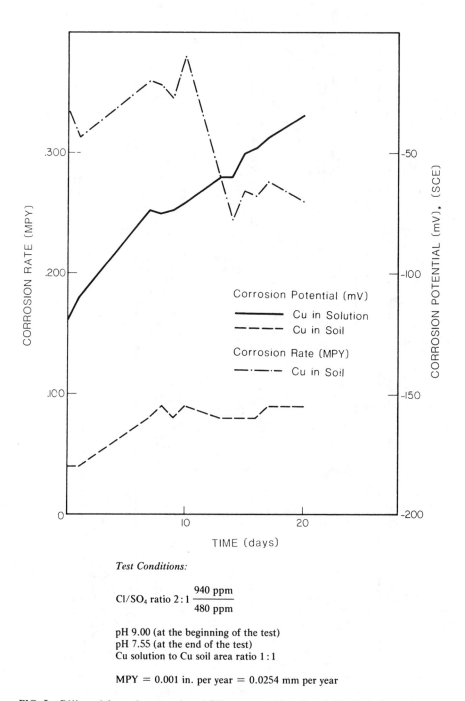

FIG. 5—*Differential aeration corrosion of CN copper wire in soil and chloride/sulfate solution.*

TABLE 1—Differential aeration effect on corrosion rate of copper CN wire in different environments.

Cathode Half Cell Environment	pH		Weight Loss Corrosion Rate, MPY					
			Free Exposure		Differential Aeration Cells			
					Cathode to Anode Ratio, 1:1		Cathode to Anode Ratio, 1:100	
	Test Start	Test End	With Aeration	Without Aeration	Aerated Solution or Backfill (Cathode)	Soil San Ramon (Anode)	Aerated Solution (Cathode)	Soil San Ramon (Anode)
Acidic Seawater, ASTM D 1141-52[a]	3.00	5.80	12.3	4.6	10.2	0.54	10.1	0.14
Seawater, ASTM D 1141-52	3.00	6.20	...	...	1.2[b]	0.20[b]	...	...
Acidic Chloride[c]	9.70	8.30	1.60	0.69	1.4	0.03	...	...
Sulfate Solution	3.00	6.90	9.9	2.9	0.03	0.3 (0.64)[d]	6.9	0.06
Chloride/sulfate[c] solution	9.00	7.55	0.69	0.33	0.11	0.62	...	...
Deionized water	7.0	6.8	0.25	0.07	0.20	0.4	...	...
Soil San Ramon	...	...	...	0.03	...	...	...	...
Imported backfill	...	...	...	...	0.005	0.33	...	...

[a] ASTM Specification for Substitute Ocean Water.
[b] Differential aeration in conduits.
[c] Chloride/sulfate ratio 2:1 (940 ppm/480 ppm).
[d] Soil contaminated with chlorides.
Note: MPY = 0.001 in. per year = 0.0254 mm per year.

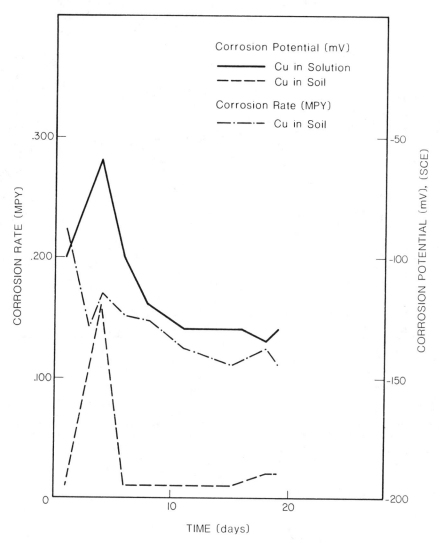

FIG. 6—*Differential aeration corrosion of CN copper wire in soil and imported backfill (Cu backfill to Cu soil area ratio 1:1).*

Besides, the copper corrosion rate in the soil can also increase, while the corrosion process is only partially controlled by differential aeration, as we found when soil was contaminated with chlorides (Figs. 7 and 8). In this case the copper anodic reaction, first controlled by oxygen reduction at the cathode (Fig. 9), is further accelerated by the chemical interaction between copper and chlorides from the soil (Figs. 7 and 8). When copper anodic reaction depletes all salt content from the soil, the passivated copper becomes only slightly affected by further copper cathodic polarization from aeration, and, therefore, the whole corrosion process can be only partially controlled by differential aeration (Fig. 10).

If a high concentration of chlorides occurred in the aerated zone (Table 1, Figs. 11 and 12), for example, due to the presence of seawater, the corrosion loss of each electrode in the couple

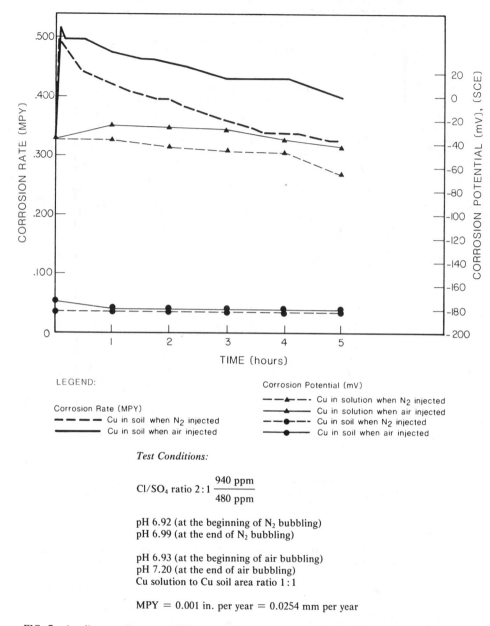

LEGEND:

Corrosion Potential (mV)

Corrosion Rate (MPY)

− − −▲− − · Cu in solution when N_2 injected
────▲──── Cu in solution when air injected

− − − − − Cu in soil when N_2 injected
───────── Cu in soil when air injected

− − −●− − · Cu in soil when N_2 injected
────●──── Cu in soil when air injected

Test Conditions:

Cl/SO$_4$ ratio 2 : 1 $\dfrac{940 \text{ ppm}}{480 \text{ ppm}}$

pH 6.92 (at the beginning of N_2 bubbling)
pH 6.99 (at the end of N_2 bubbling)

pH 6.93 (at the beginning of air bubbling)
pH 7.20 (at the end of air bubbling)
Cu solution to Cu soil area ratio 1 : 1

MPY = 0.001 in. per year = 0.0254 mm per year

FIG. 7—*Anodic corrosion rate of CN copper wire controlled by differential aeration mechanism (soil contaminated with seawater).*

was not found to be controlled by differential aeration. Under these conditions the aeration could not cathodically polarize the copper half cell surface already under attack by seawater corrosion (Figs. 11 and 12). Therefore, the copper anodic reaction in soil decreased and the real distribution of corrosion was reversed; that is, the more aerated copper specimen became the anode and corroded more than the less aerated.

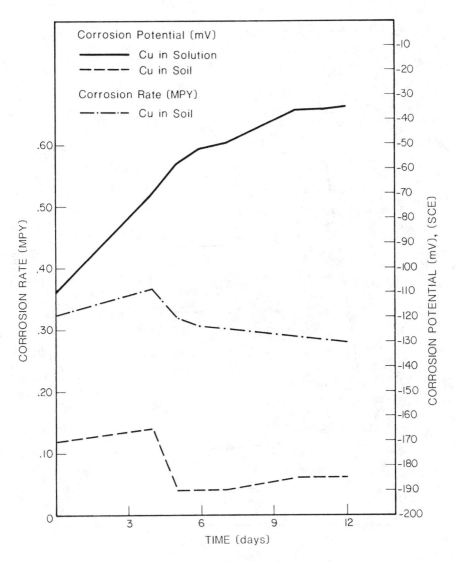

FIG. 8—*Differential aeration corrosion of CN copper wire in soil and acidic chloride/sulfate solution (soil contaminated with seawater).*

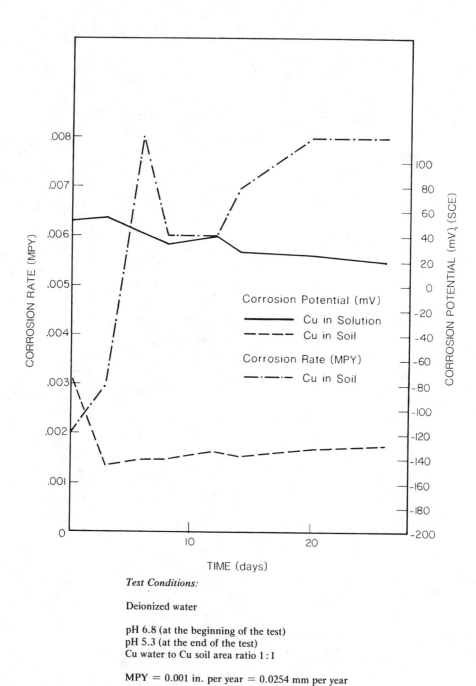

FIG. 9—*Differential aeration corrosion of CN copper wire in soil and deionized water.*

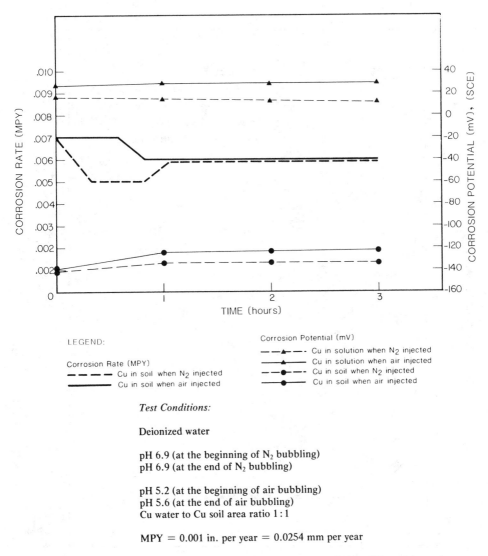

LEGEND:

Corrosion Rate (MPY)

– – – – Cu in soil when N₂ injected

——— Cu in soil when air injected

Corrosion Potential (mV)

– –▲– – Cu in solution when N₂ injected

———▲——— Cu in solution when air injected

– –●– – Cu in soil when N₂ injected

———●——— Cu in soil when air injected

Test Conditions:

Deionized water

pH 6.9 (at the beginning of N₂ bubbling)
pH 6.9 (at the end of N₂ bubbling)

pH 5.2 (at the beginning of air bubbling)
pH 5.6 (at the end of air bubbling)
Cu water to Cu soil area ratio 1 : 1

MPY = 0.001 in. per year = 0.0254 mm per year

FIG. 10—*Anodic corrosion rate of CN copper wire partially controlled by differential aeration.*

Dominant Factors of Differential Aeration

Our experiments revealed that, besides the diffusion of oxygen and the nature of the environment mentioned earlier [5–7], the cathode to anode area ratio and pH also influenced the differential aeration effect on copper corrosion in soil.

The highest corrosion rates were found at a cathode to anode area ratio of 1 : 1. Decrease of the anode (soil) area by 100 times in the presence of seawater caused an unexpected decrease of copper corrosion rate in the soil (Fig. 13). Figure 13 also shows that the corrosion rate was only partially controlled by differential aeration under these circumstances.

Previous works [8–10] indicated that the differential aeration principle can be applied only at

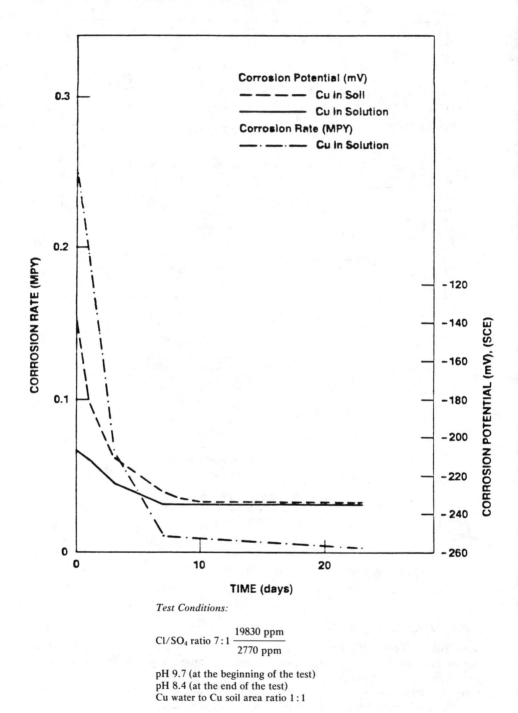

FIG. 11—*Anodic corrosion rate of CN copper wire without differential aeration control.*

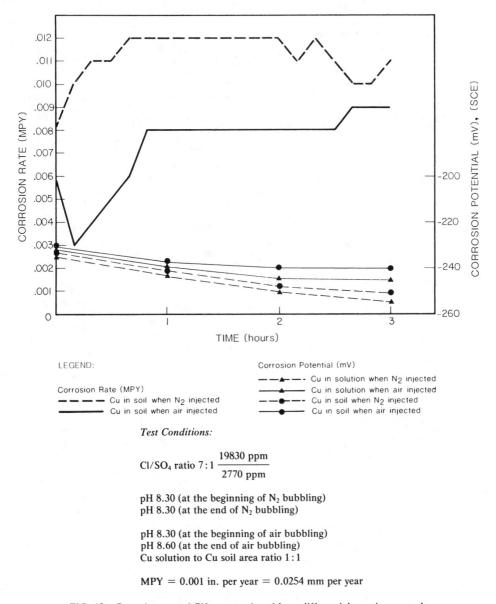

LEGEND:

Corrosion Rate (MPY)
– – – – Cu in soil when N_2 injected
——— Cu in soil when air injected

Corrosion Potential (mV)
— –▲– –· Cu in solution when N_2 injected
——▲——— Cu in solution when air injected
— –●– –· Cu in soil when N_2 injected
——●——— Cu in soil when air injected

Test Conditions:

Cl/SO$_4$ ratio 7:1 $\dfrac{19830 \text{ ppm}}{2770 \text{ ppm}}$

pH 8.30 (at the beginning of N_2 bubbling)
pH 8.30 (at the end of N_2 bubbling)

pH 8.30 (at the beginning of air bubbling)
pH 8.60 (at the end of air bubbling)
Cu solution to Cu soil area ratio 1:1

MPY = 0.001 in. per year = 0.0254 mm per year

FIG. 12—*Corrosion rate of CN copper wire without differential aeration control.*

certain pH levels and that the question of $O_2 - H^+$ concentration cells is very complicated in the presence of differential aeration. Therefore, the introduction of pH as a variable factor made our cells purposely complex. The starting pH 3 and pH 9 of our cells changed during the exposures, establishing between pH 6 to pH 8 on the cathodic (aerated solution) side. Under these conditions, H^+ ions practically could not compete with dissolved O_2 molecules because H^+ could not be reduced and cells never became pH cells. In addition to Pourbaix's conclusions [10,11] that differential aeration cells normally operate between pH 8 and pH 10, we found

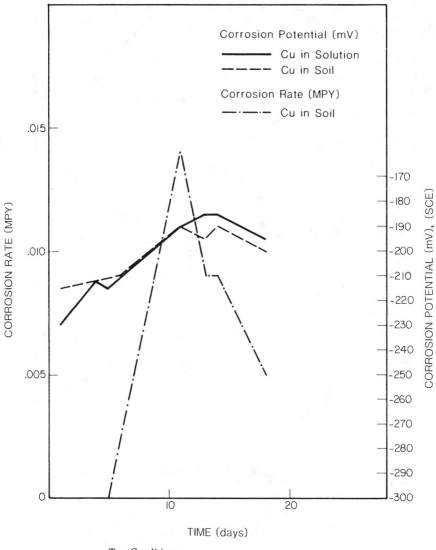

TIME (days)

Test Conditions:

Cl/SO$_4$ ratio 7:1 $\dfrac{19830 \text{ ppm}}{2770 \text{ ppm}}$

pH 3.00 (at the beginning of the test)
pH 6.20 (at the end of the test)
Cu solution to Cu soil area ratio 100:1

MPY = 0.001 in. per year = 0.0254 mm per year

FIG. 13—*Differential aeration corrosion of CN copper wire in soil and acidic seawater (ASTM D 1141-52).*

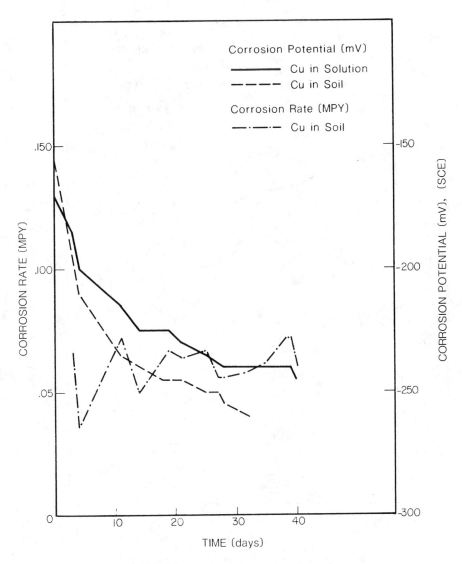

Test Conditions:

$$Cl/SO_4 \text{ ratio } 7:1 \; \frac{19830 \text{ ppm}}{2770 \text{ ppm}}$$

pH 3.00 (at the beginning of the test)
pH 5.80 (at the end of the test)
Cu solution to Cu soil area ratio 1:1

MPY = 0.001 in. per year = 0.0254 mm per year

FIG. 14—*Differential aeration corrosion of cable with CN copper wires in soil and acidic seawater (ASTM D 1141-52).*

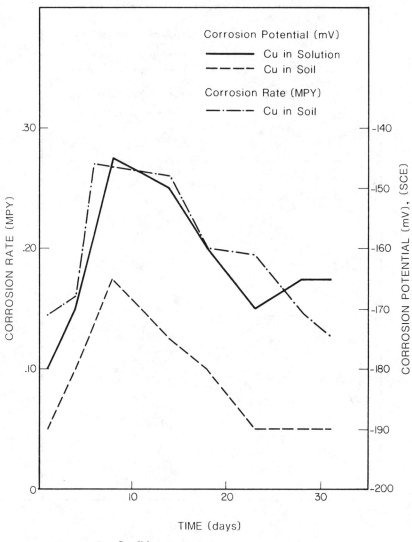

FIG. 15—*Differential aeration corrosion of CN copper wire in soil and acidic seawater (ASTM D 1141-52).*

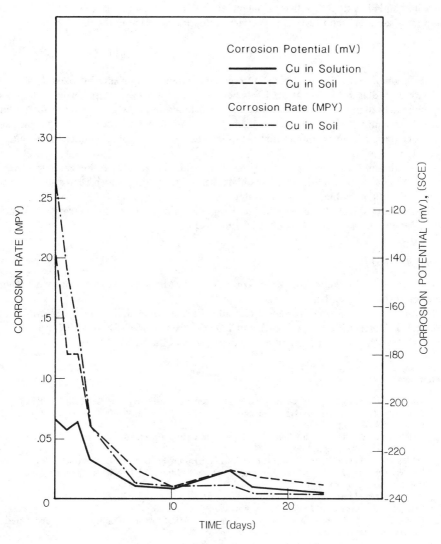

FIG. 16—*Differential aeration corrosion of CN copper wire in soil and seawater (ASTM D 1141-52).*

(Figs. 14 and 15) that differential aeration mechanism can also control the copper corrosion in the soil at a pH less than 8. When pH increased to 9 in the presence of seawater (Fig. 16), the differential aeration had practically no effect on copper corrosion in the soil.

Conclusions

Evaluation of results shows that the differential aeration mechanism controlled the CN copper wire corrosion in the soil at pH levels between 6 and 8. Copper wire corrosion increased approximately 20 times in the soil when soil was in contact with a low concentration of chlorides and sulfates and ten times when soil was in contact with imported backfill. Anodic corrosion rate of copper, caused by differential aeration, doubled when contamination from salts increased in the soils.

The maximum corrosion rate of CN wires in the soil was found when the cathode to anode area ratio was 1 : 1. Differential aeration did not control the CN copper wire corrosion in the soil when soil was in combination with seawater.

The results of this study were applied to designing methods of effective corrosion mitigation on copper concentric neutrals installed in conduits [1].

Acknowledgments

The author would like to express his appreciation to Michael Watson for his experimental work and help with graphs.

The project was cofunded by Pacific Gas and Electric Co. (PG&E) and the Electric Power Research Institute (EPRI). J. A. Hanck and T. J. Kendrew were PG&E and EPRI project managers. George Nekoksa commented on the manuscript of the EPRI report.

References

[1] "Methods for Mitigating Corrosion of Copper Concentric Neutral Wires in Conduit," EPRI EL-4981, Project 1771-1, Final Report, Department of Engineering Research, Pacific Gas and Electric Co., San Ramon, CA, January 1987.

[2] Puschel, M. A., "Corrosion—Differential Aeration Causes Problems," *Electrical World*, Vol. 182, No. 3, 1 Aug. 1974, pp. 53–54.

[3] Schick, G., "Power Cable Concentric Neutral Corrosion by Differential Aeration," presented at NACE Corrosion/78, Paper No. 52, Houston, TX, 6–10 March 1978, National Association of Corrosion Engineers, Houston, TX.

[4] Pourbaix, M., *Lectures and Electrochemical Corrosion*, 1973, pp. 220–223.

[5] Evans, U. R., *Industrial Eng., Chem.*, Vol. 17, 1925, pp. 363, 370.

[6] Evans, U. R., *An Introduction to Metallic Corrosion*, Edward Arnold Publisher, Ltd., and American Society for Metals, Metals Park, OH, 1982, pp. 35, 112, 142.

[7] Herzog, E., *Chimie et Industrie*, Vol. 27, 1932, p. 351.

[8] Shreir, L. L., *Corrosion*, Vol. 1, 1979, p. 135.

[9] Schasl, E., March, G. A., *Corrosion*, Vol. 16, 1960, p. 461t.

[10] Pourbaix, M., *Journal of the Electrochemical Society*, Vol. 101, 1954, p. 217.

[11] Pourbaix, M., "Rapports Technique," *CEBELCOR*, No. 84, 1960.

Göran Camitz[1] and Tor-Gunnar Vinka[1]

Corrosion of Steel and Metal-Coated Steel in Swedish Soils—Effects of Soil Parameters

REFERENCE: Camitz, G. and Vinka, T.-G., **"Corrosion of Steel and Metal-Coated Steel in Swedish Soils—Effects of Soil Parameters,"** *Effects of Soil Characteristics on Corrosion, ASTM STP 1013*, V. Chaker and J. D. Palmer, Eds., American Society for Testing and Materials, Philadelphia, 1989, pp. 37–53.

ABSTRACT: This report presents a systematic long-term field study of corrosion in soil of carbon steel and steel coated with zinc and an aluminium-zinc alloy (55% Al/Zn). Exposure is taking place in seven localities with different types of soil. In the report, the effects on corrosion of the groundwater table, embedment in sandfill, soil pH, type of soil, specimen size, and time of exposure are evaluated. The results are based on up to four years of exposure.

KEY WORDS: soil corrosion, effects of soil parameters, field exposure, carbon steel, zinc coating, aluminium-zinc coating, metallic coating

During a period of many thousands of years that extended until roughly 10,000 years ago, the Scandinavian peninsula was covered by an ice sheet several hundred metres thick (the Weichselian Ice Age). As a consequence of the movements of the ice, the upper layer of bedrock was scraped away; the material that was released returned later as soil. The formation of contemporary inorganic soils was started therefore during the glacial retreat and continued under the influence of water movements until long after the ice had melted away. Swedish soil is therefore rather young from a geological viewpoint.

From the corrosion point of view it is important to note that the fine-grained soils usually have a high groundwater table and that they are poorly leached out of salts. As a result of the composition of the bedrock (mostly granite and gneiss), the soils have in general a low calcium carbonate concentration and in some areas a high concentration of sulphides. Also, soils with low pH values are frequently occurring.

Previously, no systematic studies have been carried out on different metals buried in Swedish soils. In 1979, therefore, the Swedish Corrosion Institute began a long-term field study aimed at charting the corrosion of different metals buried in typical Swedish soils and the effects of different soil parameters on corrosion. This report presents the results hitherto of the exposure of steel and metal-coated steel and the effects of some soil parameters.

Experimental

Test Sites

Seven test sites in soil have been installed in different parts of Sweden (Fig. 1). The sites were chosen so that the soil at each consists of a relatively uniform type of soil representative of

[1]Swedish Corrosion Institute, Roslagsvägen 101, hus 25, S-104 05 Stockholm, Sweden.

FIG. 1—*Location of the test sites in soil.*

Swedish geology. The soils are: clay (two types), muddy clay (two types), silty clay (sulphide rich black clay), peat, and sand. The soil at the sites is naturally stratified and has never been dug over. There is no risk of stray current corrosion. The sites were also chosen so that the groundwater table in the test trenches is about 1 m below ground level in the autumn and has relatively small seasonal variations. Specimens can thus be placed above and below the groundwater table to study its effect on corrosion. Due to the great groundwater table variations, the specimens above and below the groundwater table at the Linköping test site (sand) had to be exposed at two places about 100 m apart. The pH and carbonate concentration therefore differ somewhat between the two, but otherwise the soil is similar.

Soil Analysis

The character of the soils can be seen in Table 1. The analyses were carried out on samples taken from the disturbed soil in the test trenches at a depth corresponding to the location of the metal specimens. The resistivity and pH value of the sandfills that surround some of the specimens, measured about three years after start of exposure, is shown in Table 2. All analyses given were made in the autumn of 1986.

The soil resistivity was measured using a soil box in a naturally moist soil sample. When carrying out the pH measurements, the naturally moist soil sample was shaken in deionized water at a weight ratio of 1:2.5, respectively.

TABLE 1—*Characterization of the soil at the Swedish Corrosion Institute's test sites.*

Test Sites	Depth, m	Type of Soil	Resistivity, ohm·cm	Water Content, wt% of Wet Soil	Organic Content, wt% of Dry Soil	pH Value	Carbonate, wt% CO_3^{2-} of Dry Soil	Sulphur Compounds, mg S/kg Dry Soil			Chloride, mg Cl^-/kg Dry Soil
								Sulphide, S^{2-}-S	Sulphate, SO_4^{2-}-S	Total Sulphur	
1. Enköping	0.7	Heavy clay	3 290	31	1.6	6.6	0.11	7	21	190	20
	1.7	Very heavy clay	3 450	32	0.7	6.9	0.10	7	29	150	20
2. Sollentuna	0.7	Heavy clay	3 770	41	2.9	4.3	0.19	8	202	2 090	34
	1.7	Heavy clay	1 170	48	2.2	6.3	0.19	288	526	10 300	22
3. Kramfors	0.7	Silty clay	2 570	33	0.9	6.0	0.12	238	49	2 030	50
	1.7	Silty clay	1 430	30	1.9	6.5	0.12	347	47	1 240	60
4. Gothenburg	0.7	Heavy muddy clay	1 710	41	3.7	4.4	0.10	8	412	1 480	170
	1.7	Heavy muddy clay	345	54	4.6	7.4	0.13	82	322	14 600	2 200
5. Stockholm	0.7	Heavy muddy clay	5 220	43	2.8	4.2	0.11	8	188	1 840	30
	1.7	Muddy clay	1 050	51	4.5	5.4	0.09	19	758	6 400	140
6. Laxå	0.7	Fibrous peat	7 160	85	75	4.3	<0.06	<61	36	120	180
	1.7	Pseudofibrous peat	13 100	92	61	4.2	<0.06	<31	41	550	220
7. Linköping	0.7	Sand	262 000	7	0.5	5.7	0.07	<5	6	90	20
	1.7	Gravelly sand	17 900	13	0.2	8.0	1.8	<5	8	140	20

TABLE 2—*Resistivity and pH value in the sandfills used at the test sites measured about three years after the start of exposure.*

Test Site	Depth, m	Resistivity, ohm·cm	pH Value
1. Enköping	0.1	10 300	6.5
	1.7	13 500	7.1
2. Sollentuna	0.7	18 500	6.8
	1.7	10 900	8.1
3. Kramfors	0.7	14 900	5.9
	1.7	12 000	6.7
4. Gothenburg	0.7	3 660	7.2
	1.7	1 670	8.2
5. Stockholm	0.7	30 700	5.1
	1.7	2 260	5.6
6. Laxå	0.7	56 900	6.4
	1.7	45 500	7.0

Tested Materials and Specimen Design

The materials tested and the design of specimens are given in Table 3. Two types of specimen are being used: flat bars and panels. A reason for using differently shaped specimens is to compare the influence of surface area on corrosion. Duplicate specimens are being used throughout for flat bars and triplicate specimens for panels.

The flat bar series includes 2-m-long carbon steel bars. Exposure was begun in the autumn of 1979 at five test sites. The panel series includes specimens for three materials: carbon steel, zinc-coated steel, and Aluzink-coated steel. Aluzink is the Swedish trade name for an aluminium-zinc alloy developed in the United States under the trade name Galvalume. Exposure began in 1983 except at the Kramfors test site, where it was begun in 1985.

Position of the Specimens in the Soil

The specimens were buried in trenches at the test sites—one trench for each planned intake of samples. The specimens were placed in a special pattern permitting the study of the effects of different soil parameters. In each trench there is an upper test level at about 0.7 m depth and a lower one at about 1.7 m. The groundwater table fluctuates with the season of the year between these two levels (Fig. 2).

The flat bars are exposed horizontally in the original soil above and below the groundwater table. The test panels are exposed standing in a vertical position in a row in the original soil above and below the groundwater table. Furthermore, a set of panels above and a set below the groundwater table were entirely embedded in a sandfill. The objective is to study whether embedment in a homogeneous sandfill may diminish the uniform or pitting corrosion. On refilling the trenches an attempt was made as far as possible to return the excavated soil according to its original stratification.

Evaluation of Corrosion Rate

The corrosion rate of all specimens has been determined as weight loss. Carbon steel was pickled in Clarke's solution (Sb_2O_3 20 g/L and $SnCl_2$ $2H_2O$ 60 g/L in concentrated HCl) using repeated pickling at room temperature. Zinc-coated and Aluzink-coated specimens were pickled in chromic acid solution (CrO_3 200 g/L and $BaCrO_4$ 1 g/L in deionized water) using re-

TABLE 3—Tested materials and specimen design.

Materials	Specimen Size, Length by Weight by Thickness, mm	Metal Coating as per Three-Point Test on Continuously Hot-Dipped Steel		Measured Thickness Interval of Coating, μm	Density, kg/m³
		Weight of Coating, g/m² Double-Sided	Thickness of Coating, μm		
Carbon steel SS 1312[a]	2000 by 100 by 15 (flat bars)	...	...	...	7 800
Carbon steel SS 1312	150 by 100 by 3 (panels)	...	...	...	7 800
Hot-dip galvanized steel (Piecewise hot-dipped carbon steel SS 1312)	150 by 100 by 2 (panels)	...	...	65-290[b]	7 100
Aluzink AZ 185, 55 wt% Al, 43.4 wt% Zn, 1.6 wt% Si	150 by 100 by 1.5 (panels)	185	25	...	5 000

[a]SS 1312 = Swedish standard 14 13 12, carbon steel with nominal chemical composition (wt%): C max 0.20, P max 0.050, S max 0.050, and N max 0.009. Recommended composition (wt%): Si about 0.25 and Mn 0.4 to 0.7.

[b]Mean value 164 μm, standard deviation 33 μm. The mean value corresponds to the weight of coating 1150 g/m² single sided.

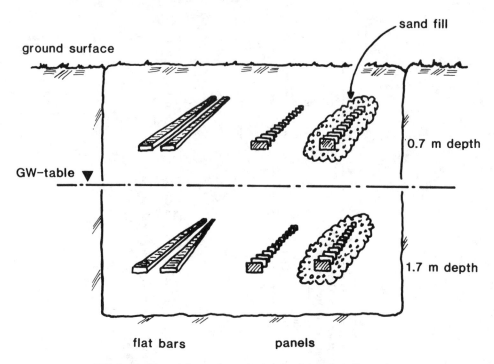

FIG. 2—*Position of the specimens in the trenches at the soil test sites.*

peated pickling at 80°C. For conversion of weight loss to average penetration, densities as in Table 3 have been used.

The Aluzink coating is a two-phase alloy consisting of about 80 vol% Al-rich phase and about 20 vol% Zn-rich phase, and under atmospheric corrosion it is the Zn-rich phase that is preferentially attacked [1]. Since the two phases have rather different corrosion properties, uncertainties may arise when evaluating the corrosion rate of the Aluzink coating. In this study it has been assumed that in both the corrosion and the pickling processes it is the Zn-rich phase which is preferably attacked. On calculating the corrosion rate, the density of the Zn-rich phase has been taken as being 5000 kg/m^3 and the weight loss has been divided by the total area of the panel. The pit depth on the steel panels has been measured with a needle instrument having a reading accuracy of 0.020 mm.

Results and Discussion

The main emphasis in the discussion has been laid on a summing up of the results, and an attempt is made to distinguish tendencies. Owing to the comparatively short exposure periods, no definite conclusions have been drawn about the effects of different soil parameters on the corrosion rates. A more detailed analysis, including the time dependence of the corrosion rate, will be performed after evaluation of specimens from prolonged exposure periods.

Corrosion rates

The corrosion rate on flat bars after four years of exposure is shown in Fig. 3. The corrosion rate on steel panels after three years of exposure expressed as average penetration is shown in

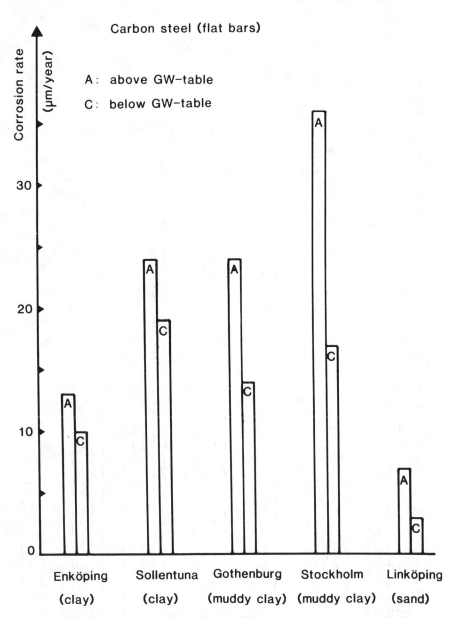

FIG. 3—*Corrosion rates on carbon steel flat bars based on about four years of exposure at the test sites in soil.*

Fig. 4 and maximum pitting rate in Fig. 5. The corrosion rate expressed as average penetration of zinc-coated panels after three years of exposure is shown in Fig. 6 and of Aluzink-coated panels in Fig. 7. Note that the exposure period for all materials at the Kramfors test site is only one year and that the absolute value of the corrosion rate can change with time. It has been considered, however, that the results from the Kramfors test site are of interest in the evaluation

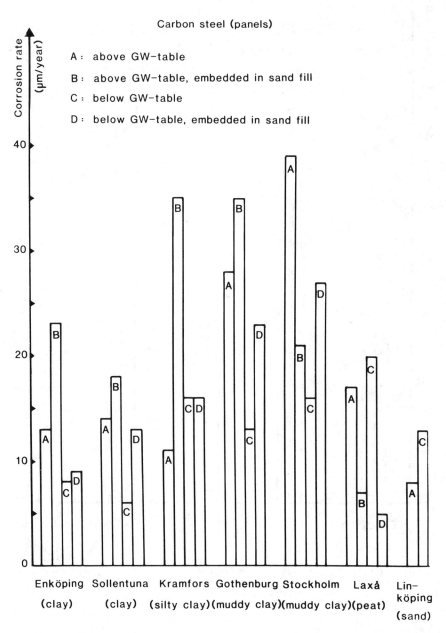

FIG. 4—*Corrosion rates on carbon steel panels based on about three years of exposure at the test sites in soil.* (Note: *only one year exposure time at the Kramfors test site.*)

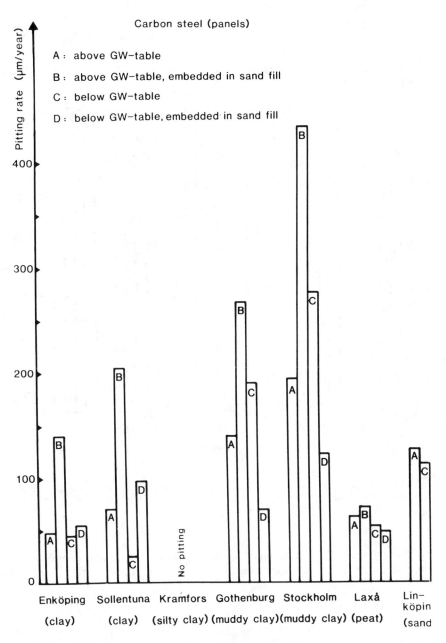

FIG. 5—*Pitting rate on carbon steel panels based on about three years of exposure at the test sites in soil.* (Note: *only one year exposure time at Kramfors test site.*)

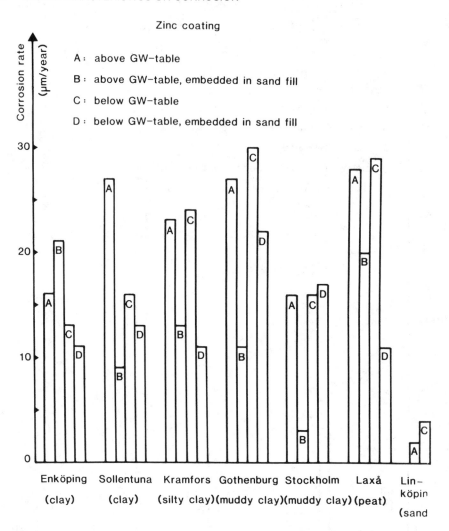

FIG. 6—*Corrosion rates on zinc-coated panels based on about three years of exposure at the test sites in soil. (*Note: *only one year exposure time at Kramfors test site.)*

of the relative influence of the position of the specimen in the trench on the corrosion. Results after one year exposure at all other test sites have been reported elsewhere [2].

Effect of the Groundwater Table

Here corrosion is compared on flat bars and panels which have been exposed above the groundwater table with corrosion on those exposed below the groundwater table. The difference in corrosion rate of specimens at these two test levels is shown in Fig. 8. At all the test sites the corrosion rate on flat bars is higher above the groundwater table than below (Fig. 8a). In most cases the corrosion rate on steel panels is considerably higher above than below the groundwater table (Fig. 8b).

Aluzink coating

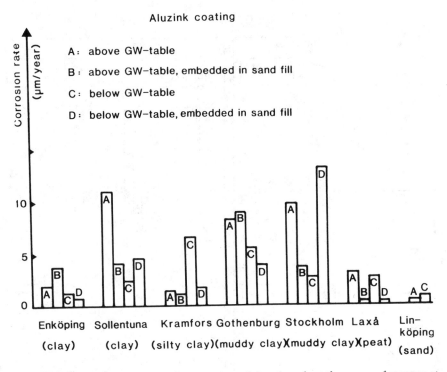

A: above GW-table

B: above GW-table, embedded in sand fill

C: below GW-table

D: below GW-table, embedded in sand fill

FIG. 7—*Corrosion rates on Aluzink-coated panels based on about three years of exposure at the test sites in soil. (Note: only one year exposure time at Kramfors test site.)*

The reason for the higher corrosion rate at the upper test level can be that the aeration of the soil is better above the groundwater table. But, however, the lower pH value of the soil above the groundwater table can also have been a contributing factor. The groundwater table acts as a barrier to the transportation of oxygen down through the soil. In soil, the transportation of oxygen takes place primarily by diffusion [3,4]. Above the groundwater table the pores in the soil are mostly air-filled, which means that oxygen can be transported by diffusion in the gas phase. This gives rise to faster transportation of oxygen to the steel surfaces situated at the upper test level, which in turn can stimulate the cathode process and thereby increase the corrosion attack.

But the pores below the groundwater table are, on the other hand, completely or almost completely water-filled, which means that oxygen must be primarily transported by diffusion in the aqueous phase. This gives rise to a considerably slower transportation of oxygen, which can reduce the cathode process and thereby decrease the corrosion rate of the steel specimens at the lower test level. Since the water content of the soil and consequently the degree of water saturation above the groundwater table is dependent on the season of the year, the greatest difference in aeration between the test levels above and below the groundwater table is during the dry seasons.

The pitting rate on steel panels in the original soil varies unsystematically in relation to the groundwater table from test site to test site (Fig. 5). However, in sandfills it is considerably higher above the groundwater table than below. The corrosion rates on zinc-coated and Aluzink-coated panels show no clear tendency with respect to their position related to the groundwater table (Figs. 8c and 8d).

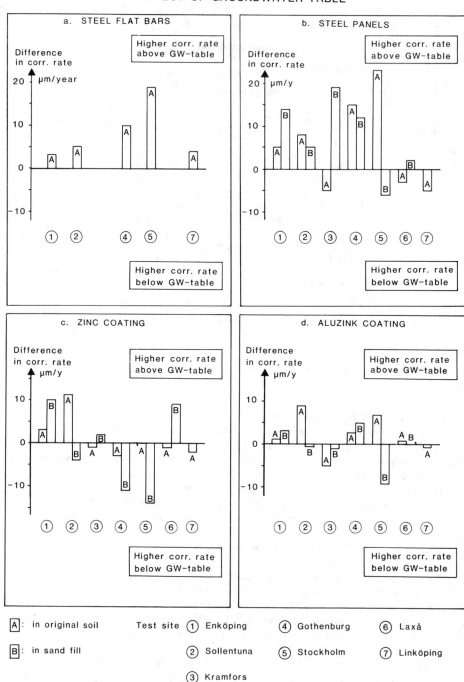

FIG. 8—*Effect of groundwater table (GW table) on the corrosion of steel flat bars, steel panels, and zinc-and Aluzink-coated panels at each test site. The effect is expressed as the difference (D) in corrosion rate (w) calculated as D = w (above GW) − w (below GW), all based on corrosion rates in Figs. 3, 4, 6, and 7.*

Effect of Sandfill

The results permit a comparison to be made between the corrosion on panels embedded in a sandfill with corrosion on panels placed nearby but in the original soil. The difference in corrosion rate of panels in these two positions is shown in Fig. 9. The corrosion rate is higher on steel panels embedded in the sandfill than on those placed directly in the original soil at all the test sites except Laxå and Stockholm—above the groundwater table (Fig. 9a). The maximum pitting rate is highest at all the test sites on the steel panels embedded in the sandfill above the groundwater table (Fig. 5). The sandfill is considered not to have had any generally reducing effect on the corrosion on the steel panels with respect to either uniform or pitting corrosion.

In all cases except two, the corrosion rate is lower on zinc-coated panels embedded in a sandfill than on panels placed directly in the original soil nearby (Fig. 9b). The reduction in corrosion rate in the sandfill is especially great in those cases where the corrosion rate is high on nonembedded panels (Fig. 6). In most cases the corrosion rate is somewhat lower on Aluzink-coated panels embedded in a sandfill than on panels placed directly in the soil (Fig. 9c).

The influence of sandfill on steel, on the one hand, and on zinc and Aluzink coatings, on the other, appears to be different. The higher corrosion rate of steel panels in the sandfills compared to that on nearby panels in the original soil could be associated with the greater permeability of the sandfills, which facilitates the transportation of oxygen by soil-air and by water to the steel surfaces. This can stimulate the cathode process and thereby increase the corrosion rate. The two exceptions (Stockholm—upper test level, and Laxå—upper and lower test level) could be explained by the sandfill resistivity being very high in these cases and that it differs markedly from that in the nearby original soil.

The lower corrosion rate of zinc in the sandfills could depend on more favorable conditions for formation of a protective layer consisting of corrosion products on the zinc surface. Such a protective layer may reduce the corrosion rate in the sandfill compared to that in the nearby original soil. Heim [5] has shown that better aeration of the soil promotes the formation of a protective layer on a zinc surface. Grauer et al [6, 7] have shown how the combination of pH and the total concentration of carbonate in an aqueous solution play a part in the formation of a carbonate-containing protective layer. In this case the combination of a somewhat higher pH and better access to soil-air and thus carbon dioxide, which forms carbonate, can have facilitated the formation of a dense and covering layer of corrosion products on the zinc surface in the sandfills.

Effect of the pH Value in the Soil

The corrosion rate is in general higher in soils having a low pH, both on carbon steel panels and zinc-coated panels (Fig. 10). This might be explained, especially with regards to zinc, by the fact that the formation of a protective layer of corrosion products is counteracted at low pH values [7]. For both metals, the hydrogen evolution as a cathode reaction at low pH values may have contributed to the cathode process and thereby increased the corrosion.

Effect of Type of Soil

The corrosion rate on carbon steel in the original soil is lowest in sand and highest in muddy clay (Figs. 3 and 4). On zinc coatings it is lowest in sand and highest in muddy clay and peat (Fig. 6). On Aluzink coatings it is lowest in sand and highest in clay and muddy clay (Fig. 7).

Corrosion Rate of Carbon Steel as a Function of Exposure Time

Figure 11 shows how the corrosion on steel flat bars and steel panels has developed with exposure time. However, due to the small number of observations no strict mathematical curve

EFFECT OF EMBEDDING IN SAND FILL

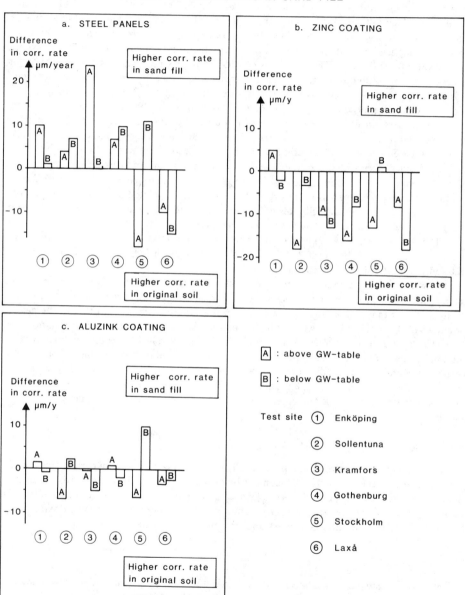

FIG. 9—*Effect of embedment in sandfill on the corrosion of steel panels and zinc- and Aluzink-coated panels at each test site. The effect is expressed as the difference (D) in corrosion rate (w) calculated as* D = w *(in sand fill)* − w *(in original soil), all based on corrosion rates in Figs. 3, 4, 6, and 7.*

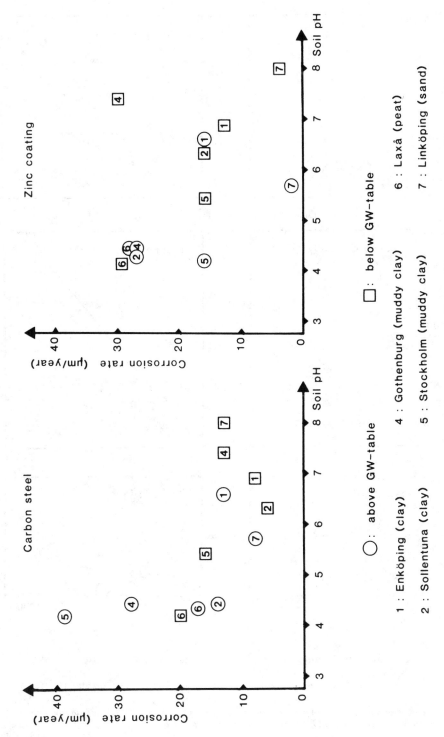

FIG. 10—Corrosion rates on carbon steel (left) and on zinc coating (right) versus soil pH value. Three years of exposure at different test sites in soil.

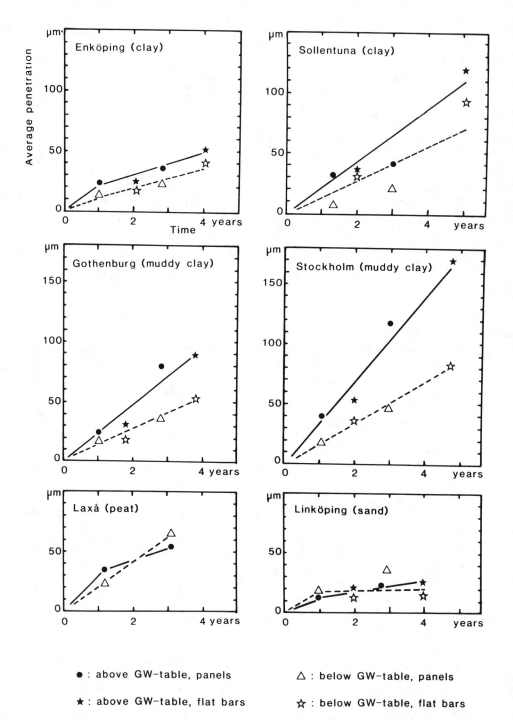

FIG. 11—*Development of corrosion on carbon steel with exposure time at six test sites in soil. Solid line: above groundwater table. Dashed line: below groundwater table.*

fitting has been done. The corrosion rates so far appear in most cases to be relatively constant with time. Furthermore, the corrosion rate is approximately the same on flat bars and steel panels, despite a very large difference in the surface area of the specimens.

Summary and Conclusions

The following conclusions may be drawn from the results obtained hitherto:

1. The corrosion rate on carbon steel specimens is higher above the groundwater table than below. On zinc coatings and Aluzink coatings the groundwater table has no distinct effect on the corrosion rate.

2. The corrosion rate is higher on carbon steel panels embedded in a homogeneous sandfill than on panels placed nearby directly in the original soil. On zinc-coated panels the corrosion rate is lower in the sandfill. On Aluzink-coated panels there is a similar tendency but not as significant as on zinc coatings.

3. The pitting rate on carbon steel panels is markedly highest in the sandfill above the groundwater table at all test sites.

4. The corrosion rate is in general higher in soils with low than in soils with high pH values on both carbon steel and on zinc coatings.

5. Muddy clay and peat have the highest corrosivity towards all three materials tested, whereas sand has the lowest.

6. The corrosion rate is approximately the same on steel panels as it is on flat bars, which have 15 times larger exposed area.

7. The corrosion rates on carbon steel so far appear in most cases to be relatively constant with time.

References

[1] Zoccola, J. C., Townsend, H. E., Borzillo, A. R., and Horton, J. B. in *Atmospheric Factors Affecting the Corrosion of Engineering Metals, ASTM STP 646,* S. K. Coburn, Ed., American Society for Testing and Materials, Philadelphia, 1978, pp. 165–184.

[2] Camitz, G. and Vinka, T.-G., "Proceedings, 10th Scandinavian Corrosion Congress," Stockholm, 1986, Bulletin No. 101, Swedish Corrosion Institute, Stockholm, 1986. pp. 305–312.

[3] Tomashov, N. D., *Theory of Corrosion and Protection of Metals,* MacMillan, New York, 1966, pp. 413–418.

[4] Wranglén, G., *An Introduction to Corrosion and Protection of Metals,* Chapman and Hall, London, 1985, pp. 125–127.

[5] Heim, G., *Technische Uberwachung,* Vol. 18, 1977, pp. 257–261 (in German).

[6] Grauer, R., *Werkstoffe und Korrosion,* Vol. 31, 1980, pp 837–850 (in German).

[7] Grauer, R. and Feitknecht, W., *Corrosion Science,* Vol. 7, 1967, pp. 629–644 (in German).

Robert C. Rabeler[1]

Soil Corrosion Evaluation of Screw Anchors

REFERENCE: Rabeler, R. C., **"Soil Corrosion Evaluation of Screw Anchors,"** *Effects of Soil Characteristics on Corrosion, ASTM STP 1013,* V. Chaker and J. D. Palmer, Eds., American Society for Testing and Materials, Philadelphia, 1989, pp. 54–80.

ABSTRACT: A corrosion evaluation study was performed in northern Minnesota to evaluate corrosion of galvanized screw anchors used as guy anchors to support a 500-kV electrical transmission line. The initial evaluation used field four-pin resistivity testing to locate areas of low resistivity. Laboratory chemical tests were performed on soil samples from selected areas. Instantaneous corrosion rates were evaluated using polarization testing techniques. Test anchors were later installed providing actual thickness and weight loss measurements.

KEY WORDS: underground corrosion, polarization, screw anchors, buried metals, foundations, transmission towers

This paper describes an evaluation performed over a seven-year period to evaluate corrosion of guy anchors for a transmission power line.

Project Description

Northern States Power (NSP), an electric utility based in Minneapolis, constructed a 500-kV power line as a joint venture with Minnesota Power and Light. The power line was extended from Minneapolis to the Canadian border. The Manitoba Hydroelectric Board extended the line from the Canadian border to Winnipeg. In all, over 650 km of transmission line were constructed. The transmission line allows for sharing of electrical power between the United States and Canada.

The portion of the line for this study consisted of approximately 800 towers in northern Minnesota. A variety of foundation and guy anchor systems were used, including steel H-piles, cast-in-place concrete piers, dome anchors, steel grillages, and galvanized screw anchors.

The galvanized screw anchors were mainly studied due to their thin cross section. Figure 1 shows a typical guy assembly.

The screw anchors were coated with zinc prior to installation according to ASTM Specification for Zinc Coating (Hot-Dip) on Iron and Steel Hardware (A 153). This required an average thickness of about 3.4 mils (1 mil = 0.001 in. = 0.00254 cm). The actual zinc thickness values measured during this study were typically greater than 3.4 mils. The underlying steel was A-4130.

Within a few months after installation in 1979, some of the guy anchor assemblies failed at the weld of the anchor couplings. At that time, questions were raised by NSP regarding possible corrosion of the foundation systems due to the adjacent soil. The questions raised initiated the corrosion evaluation study.

[1]Senior associate, Soil and Materials Engineers, Lansing, MI 48906.

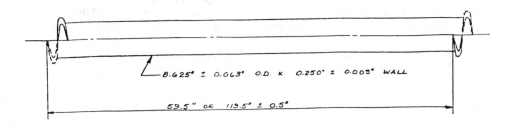

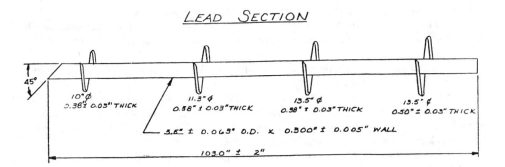

FIG. 1—*Typical anchor assembly.*

Literature

A review of the literature available at the time indicated several organizations had developed their own test procedures to evaluate corrosion. One of the best references, by Romanoff [1], was published by the National Bureau of Standards (NBS). This reference presents long-term corrosion rate data for various metals in a variety of soil conditions. However, the specimens were buried relatively shallow, and the data may not be indicative of deeply embedded specimens such as the driven piling or screw anchors used for our project.

Several references indicated polarization testing could evaluate in-place instantaneous corrosion rate [2–13]. Schwerdtfeger [3] used the polarization break technique for calculating corrosion rate of metals from polarization data. Current-potential curves are plotted on a semi-log scale (log of current versus linear potential) to evaluate the break in the anodic and cathodic curves.

The polarization resistance or linear polarization method was proposed by Stern [12] based on data of Skold and Larson [13]. This method utilizes the initial linear portion of the polarization curve (polarization current that changes less than 10 to 15 mA) to calculate corrosion current.

Both the linear polarization and polarization break techniques calculate corrosion current. Faraday's law must then be used to convert from corrosion current to corrosion rate.

Test Procedures

The first step was to determine, by four-pin resistivity testing, the location of low-resistivity soils. Approximately 10% of the towers were evaluated in this manner using the Wenner four-pin test [ASTM Method for Field Measurement of Soil Resistivity Using the Wenner Four-Electrode Method (G57-78)]. Based on this data, soil borings were performed at selected tower locations. Samples were returned to the laboratory for classification and visual examination. Resistivity tests were performed in a Miller soil box using a procedure similar to ASTM G 57-78.

To evaluate the chemical characteristics of the soil, a soil-distilled water slurry was obtained by mixing 50 g of soil and 100 mL of distilled water. Conductivity and pH tests were conducted on this slurry. The concentrations of soluble materials were evaluated from the slurry filtrate. These results were presented in units of milligram equivalents per 100 g of soil.

Water samples were also obtained from within the hollow foundation anchors underneath some of the towers. Chemical analyses were performed with reporting units of milligram equivalents per liter.

The soil types encountered in the borings are shown in Figs. 2 through 4. The laboratory test results are shown on Tables 1 through 3.

To evaluate instantaneous corrosion rates, polarization testing equipment was developed specifically for this project. Although laboratory-type equipment was available, it could not be used in the field. Considerable assistance was obtained from NBS in developing the testing procedure. We elected to modify the design presented by Schwerdtfeger [3].

To obtain accurate corrosion rate data, one must eliminate the IR drop between the specimen and the reference cell. In the test apparatus by Schwerdtfeger, a bridge circuit was used whereby the internal resistance on half of the bridge equals the soil resistance. This, in effect, nullifies the IR drop. The difficulty in using this technique is in properly balancing the resistance. Another technique to eliminate the IR drop is to use the instantaneous-off procedure. When the polarization current is turned off for an instant, the resistance immediately goes to zero, whereas depolarization is a time-dependent phenomenon. Therefore, if the potential is read immediately after the current is interrupted, the potential reading would be without the IR drop. The difficulty with this procedure is reading the polarization current at the instant the current is interrupted. If depolarization occurs very rapidly, it is difficult to read the potential.

For this project, the potential change was measured using a portable strip chart voltage recorder. This instrument had two recording pens. One pen was set to the 0 to 10-mV scale to evaluate the initial small potential changes. The other pen was set to a 0 to 100-mV scale to measure major potential changes. Change in current was measured using a Miller multicombination meter. Power was supplied using a 12-V d-c battery. During the initial portions of polarization tests, only small current flows are desired. Therefore, a voltage control circuit was used during the early stages of the tests to reduce the driving voltage and thereby reduce the current flow. Rheostats were also used to control the current flow.

Polarization tests were performed on the existing anchors following temporary electrical isolation from the towers. The counter (auxiliary) electrode consisted of other existing screw an-

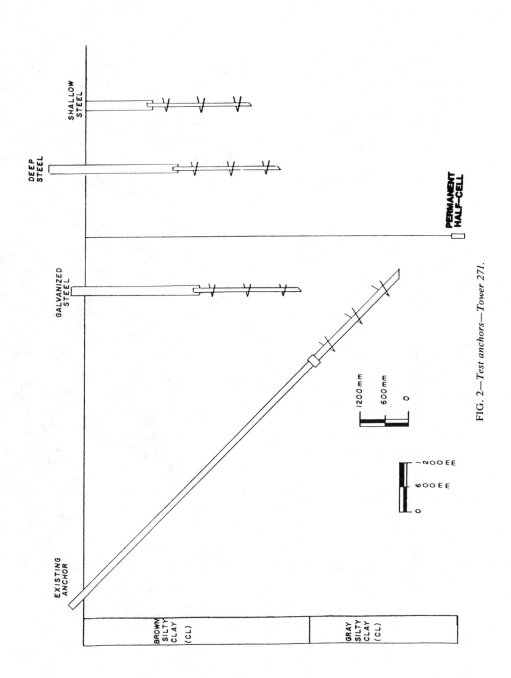

FIG. 2—Test anchors—Tower 271.

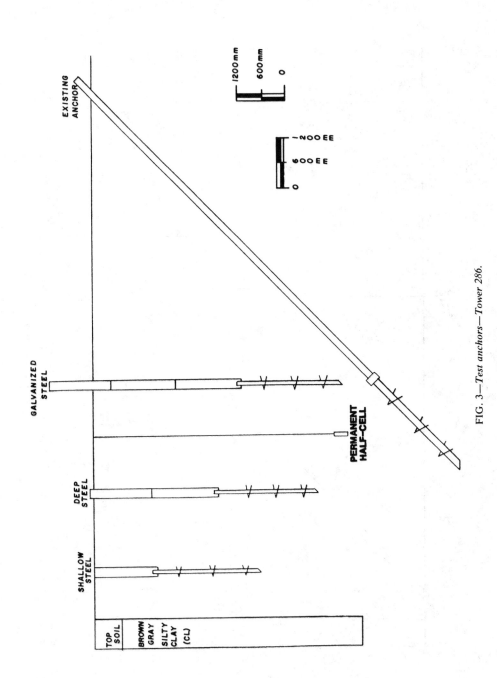

FIG. 3—*Test anchors—Tower 286.*

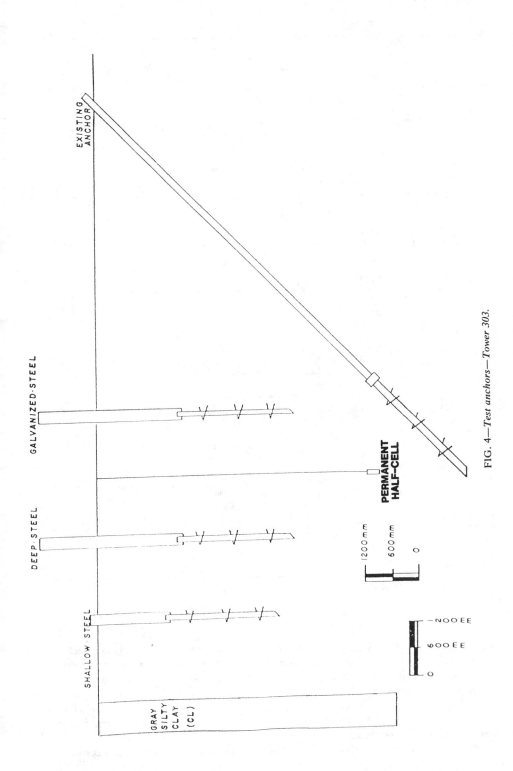

FIG. 4—*Test anchors—Tower 303.*

TABLE 1—*Summary of soil/water tests: Tower 271.*

FOUR PIN RESISTIVITY

Pin Spacing, ft	Apparent Resistivity, ohm-cm	Barnes Layer Resistivity, ohm-cm
5	2200	2200
10	2490	2860
20	1920	1560
30	2010	2230

SOIL BOX RESISTIVITY

Depth, ft	Resistivity, ohm-cm
10 to 11.5	1300
20 to 21.5	720
28 to 30	710

LABORATORY pH

Depth, ft	pH
5 to 7	8.6
10 to 11.5	8.6
15 to 17	8.3
20 to 21.5	8.0
25 to 27.5	7.9
28 to 30	7.9

CHEMICAL TESTS

Soluble Material	Water Sample, mg-eq/L	Concentration of Soluble Materials at Depth, mg-eq/100 g Soil			
		5 to 7 ft	10 to 11.5 ft	15 to 17 ft	28 to 30 ft
Chloride	0.35	0.13	0.04	0.04	0.04
Sulfate	0.98	0.02	0.02	0.06	0.09
Sodium	0.22	0.04	0.04	0.06	0.10
Alkalinity	. . .	0.22	0.23	0.12	0.39
Calcium	2.10	0.16	0.10	0.13	0.38
Nitrate	. . .	. . .	<0.01	. . .	<0.01
Sulfite	. . .	. . .	<0.01	. . .	<0.01
Potassium	. . .	. . .	<0.01	. . .	<0.01
Magnesium	0.56	. . .	0.10	. . .	0.31

NOTE: 1 ft = 0.3048 m.

chors of the tower. Tests were performed with the copper–copper sulfate reference cell at the ground surface. Results of the initial phase of the study indicated high corrosion rates were occurring. The polarization test results indicated that if the rates were to continue, the 50-year life would not be obtained. The 50-year life was assumed to occur when 30% of the cross-sectional area was lost. We were also concerned about the high sulfate content encountered, particularly at Tower 286. High sulfate concentrations in soil have been known to cause severe corrosion of ferrous and zinc metals.

At this point, the cost of cathodic protection for the anchors was evaluated. It was anticipated

TABLE 2—*Summary of soil/water tests: Tower 286.*

Four Pin Resistivity

Pin Spacing, ft	Apparent Resistivity, ohm-cm	Barnes Layer Resistivity, ohm-cm
5	1530	1530
10	1050	800
15	1090	1180
20	1190	1610

Soil Box Resistivity

Depth, ft	Resistivity, ohm-cm
3.5 to 5	620
13.5 to 15	620
18.5 to 20.5	490

Laboratory pH

Depth, ft	pH
3.5 to 5	8.4
13.5 to 15	7.7
18.5 to 20.5	8.0

Chemical Tests

Soluble Material	Water Sample, mg-eq/L	Concentration of Soluble Materials at Depth, mg-eq/100 g Soil	
		13.5 to 15 ft	18.5 to 20.5 ft
Chloride	0.08	0.04	0.05
Sulfate	0.96	6.06	1.22
Sodium	0.25	0.12	0.13
Alkalinity	...	0.19	0.36
Calcium	0.74	1.65	0.52
Nitrate	...	<0.01	...
Sulfite	...	<0.01	...
Potassium	...	<0.01	...
Magnesium	0.78	2.10	...

NOTE: 1 ft = 0.3048 m.

an impressed current, deep anode bed type system could be used. However, this would cost several million dollars. A review of the literature indicated, in many cases, corrosion rates decrease with time. Since the corrosion rate data indicated that failure would likely not occur within five years, we convinced the staff of NSP to extend the corrosion study several years to evaluate whether the corrosion rate would decrease with time.

Test Anchors

To assist in the additional evaluation, test anchors were installed at three sites (Towers 271, 286, and 303). At each site, we elected to install galvanized steel and bare steel anchors. The galvanized steel was installed to evaluate the corrosion rate of the zinc, and the steel was in-

TABLE 3—*Summary of soil/water tests: Tower 303.*

Four Pin Resistivity

Pin Spacing, ft	Apparent Resistivity, ohm-cm	Barnes Layer Resistivity, ohm-cm
5	1820	1820
10	1050	740
15	980	850
20	1000	1060

Soil Box Resistivity

Depth, ft	Resistivity, ohm-cm
13.5 to 15	580
20 to 21.5	540

Laboratory pH

Depth, ft	pH
8.5 to 10	8.3
13.5 to 15	7.9
20 to 21.5	8.1

Chemical Tests

Material	Water Sample, mg-eq/L
Chloride	0.18
Sulfate	0.13
Sodium	0.23
Calcium	0.42
Magnesium	0.71

stalled to evaluate the corrosion rate of the underlying steel. For the steel, we installed both shallow steel (SS) and deep steel (DS) anchors. The idea was to evaluate the effect of corrosion rate with depth, if possible. Additionally, since the anchors came in sections, this allowed for evaluation of corrosion rates at different levels. Figures 2 through 4 show the test anchor assemblies. The test anchors were isolated from the existing anchors. Steel corrosometer probes were installed inside the hollow portions of the test anchors to evaluate the corrosion environment inside the anchors. These probes were commercially available. The corrosometer probes consist of a steel specimen, which changes in electrical resistance when it corrodes. The resistance change can then be correlated to corrosion rate. The corrosometer probes were inserted into slotted PVC pipe so they would not contact the metal of the test anchors.

Prior to installing the test anchors, initial measurements were obtained on each section. Total section thicknesses were measured with a digital micrometer. For the galvanized anchors, the thickness of the zinc was measured using an Accuderm thickness meter. This meter induces a magnetic field. By placing a special probe on the surface of the galvanized steel, the thickness of the zinc was determined by the position of the probe in the magnetic field. Additionally, each section was weighed using a platform scale accurate to the nearest ounce. Each of the galvanized bolts connecting the sections was weighed with a triple beam gram balance.

After approximately 4.3 years, the anchors were removed from the test locations. First, each section was steam cleaned. This was followed by light sandblasting. Extreme care was taken not to remove metal while sandblasting. The anchors were then reweighed and remeasured. Weight losses and corrosion rates were then calculated. Additionally, pit depth measurements were obtained using a special dial indicator with a needle point.

While the test anchors were in the ground, polarization tests were performed. Reference cells were used at the ground surface, as discussed previously. Additionally, we installed permanent reference cells at each test anchor location near the bottom of the anchors. Very little difference in the polarization curves resulted from the different reference cell locations.

An example of polarization curves generated from Tower 286 are shown in Figs. 5 through 8. From this data, corrosion rates were calculated using both the linear polarization and polarization break techniques. These data were plotted with time and are shown in Figs. 9 through 11.

With the test anchors and polarization data, actual hard data of corrosion rates at the site were available for 4.3 years. A summary of the weight losses is shown on Table 4.

The hollow screw anchors are subject to corrosion on both the inside and outside. The outside surface is in contact with the soil; the inside surface often filled up with groundwater. The thickness losses indicated in Table 4 are for total thickness loss due to corrosion on both the inside and outside surfaces.

Discussion of Test Results

The results of the soil resistivity and chemical test data were evaluated by comparing the soil properties with those at the NBS test sites, where corrosion rate information was available [1]. This analysis indicated that relatively high corrosion rates should be expected. However, these data are for buried specimens. Slower corrosion rates would be expected for the anchors as installation would not introduce oxygen into the soil like an excavation would. Therefore, the results of the polarization tests and the test anchor data were relied on more heavily in evaluating corrosion rates.

Final weight measurements of the galvanized steel bolts indicated weight losses ranging from 0.1 to 0.2% of the original weight. For the bare steel bolts, the change in weight ranged from 0.4 to 0.7%. The bolts were in place for approximately 4.3 years.

For the galvanized test anchors, the various test methods indicated loss in total zinc thickness (inside and outside surfaces) of approximately 0.5 to 2.0 mils after 4.3 years. This corresponds to an annual average corrosion rate (on one surface) ranging from 0.06 to 0.23 mils per year (mpy). The measured corrosion rates were slightly higher at Tower 286. This could be the result of the higher sulfate concentrations at this location.

For the bare steel, the total thickness loss (inside and outside surfaces) ranged from approximately 1 to 4 mils after 4.3 years. This would correspond to average annual corrosion rates (on one surface) ranging from approximately 0.12 to 0.47 mpy.

For both the galvanized and bare steel, it appeared the polarization test measurements on the test anchors accurately predicted the actual measured corrosion rates. The polarization break technique appeared to most accurately reflect actual weight loss measurements.

Based on the data obtained from this study, there does not appear to be a clear trend with regard to corrosion rate and depth. For the galvanized steel test anchors, there was a slight trend toward an increase in corrosion rate with depth. For the bare steel test anchors, there was a slight trend towards a decrease in corrosion rate with depth.

In observing the test anchors, we noticed that the bare steel was much more severely pitted near the ground surface than the galvanized steel. The depths of the pits were not necessarily greater, but there was a much greater concentration of pits. Pit-depth measurements were obtained on the bare steel anchors. The maximum pit depths were generally on the order of 10 to 20 mils. The test anchors at Tower 286 indicated slightly deeper pit depths, with a maximum pit depth at Tower 286 of 29 mils.

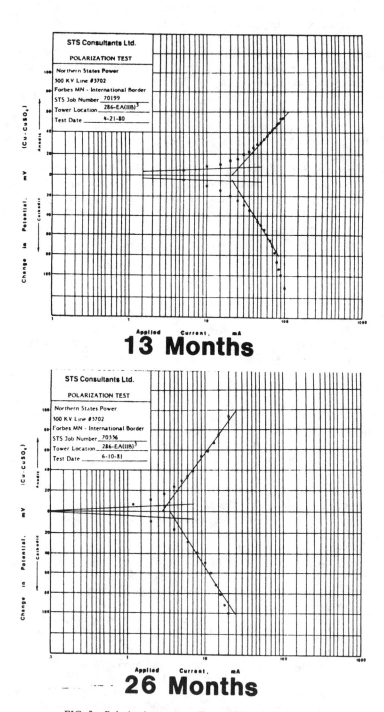

FIG. 5—*Polarization curves—Tower 286, existing anchor.*

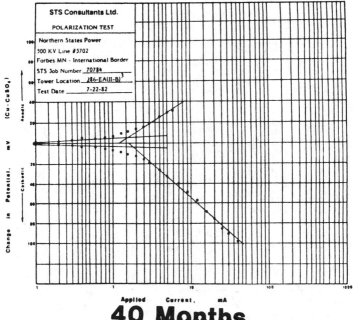

40 Months

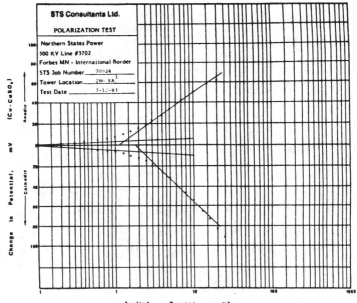

51 Months

FIG. 5—*Continued.*

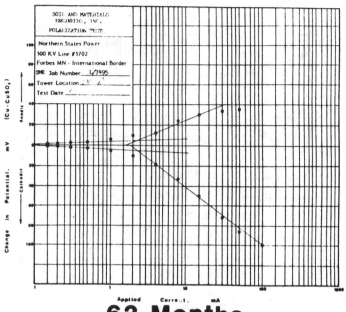

63 Months

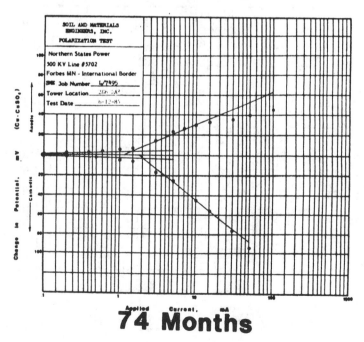

74 Months

FIG. 5—*Continued.*

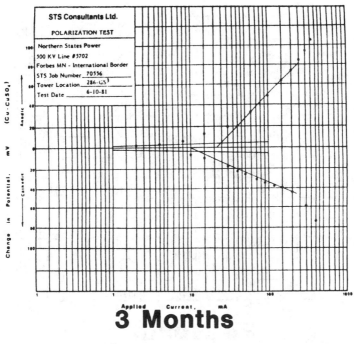

3 Months

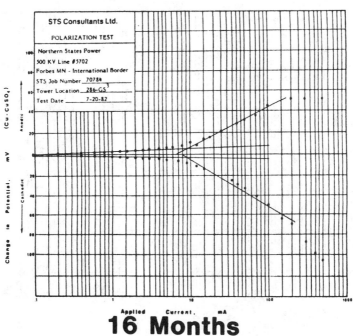

16 Months

FIG. 6—*Polarization curves—Tower 286, galvanized steel anchor.*

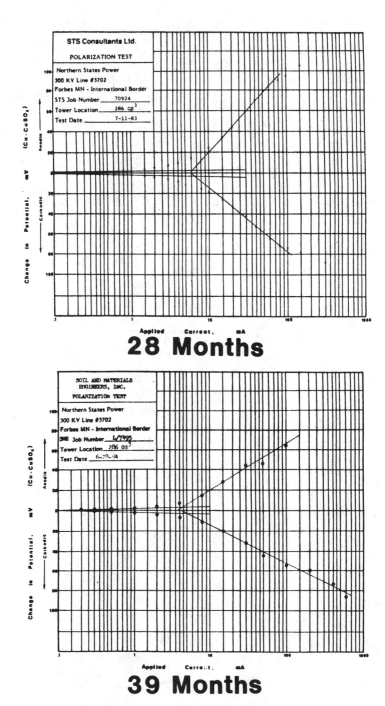

FIG. 6—*Continued.*

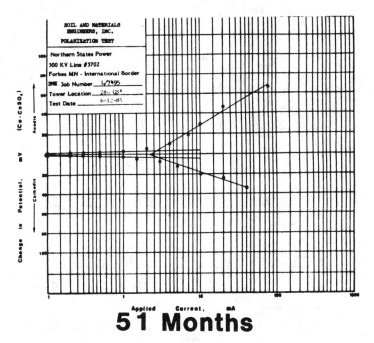

51 Months

FIG. 6—*Continued.*

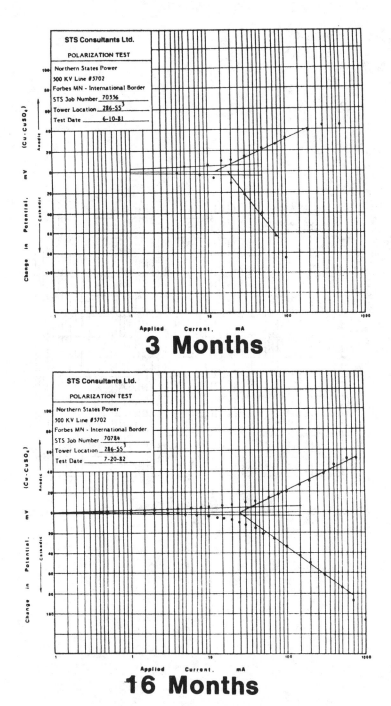

FIG. 7—*Polarization curves—Tower 286, shallow steel anchor.*

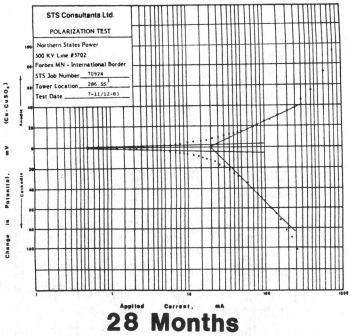

28 Months

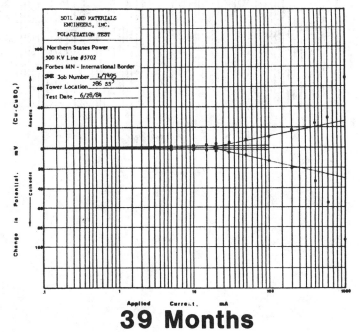

39 Months

FIG. 7—*Continued.*

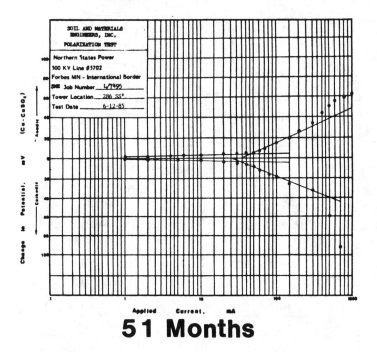

51 Months

FIG. 7—*Continued.*

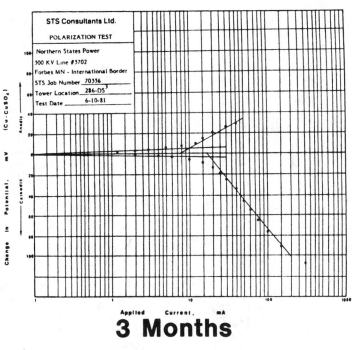

3 Months

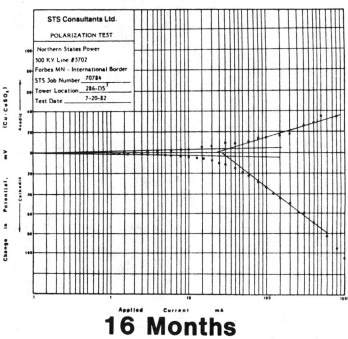

16 Months

FIG. 8—*Polarization curves—Tower 286, deep steel anchor.*

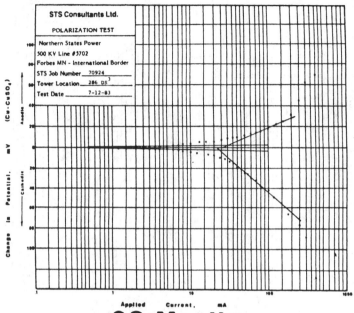

28 Months

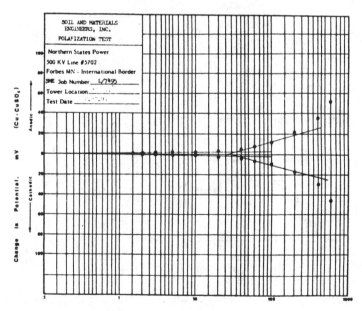

39 Months

FIG. 8—*Continued.*

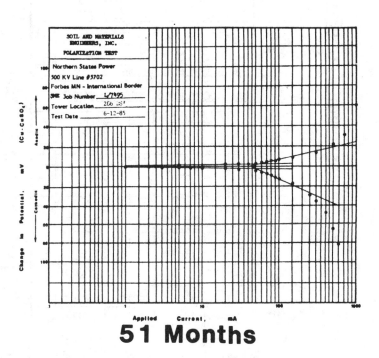

51 Months

FIG. 8—*Continued.*

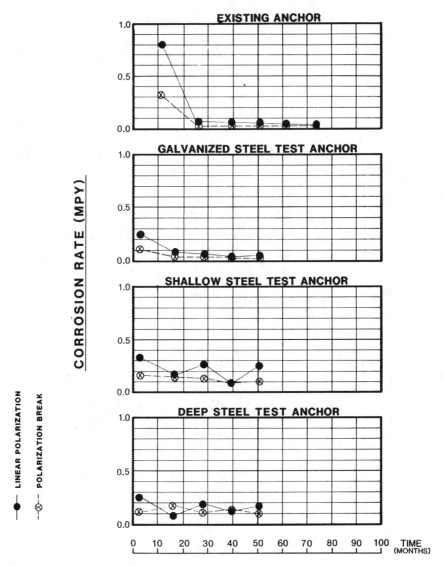

FIG. 9—*Corrosion rate versus time polarization data—Tower 271.*

Conclusions

This study has confirmed previous references which indicate polarization tests can be performed to accurately predict corrosion rates of buried metallic structures. However, corrosion measurements must be taken over a period of time to establish long-term corrosion rates.

The structural engineer for the project used a failure criteria of a loss in cross-sectional area of less than 30% due to general corrosion. Using the worst case condition for corrosion rate of the zinc, combined with the worst case condition for corrosion rate of the underlying steel, we now predict the design life will be in excess of the desired 50 years.

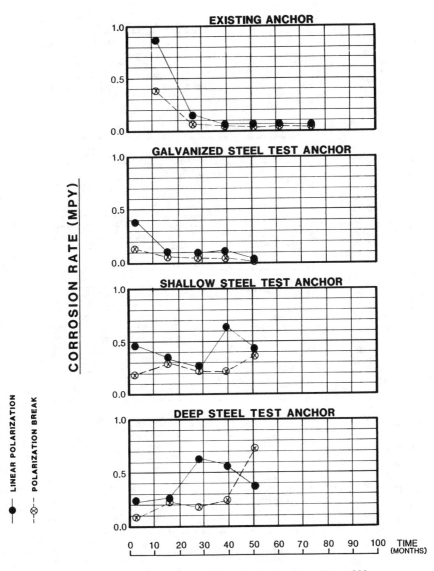

FIG. 10—*Corrosion rate versus time polarization data—Tower 286.*

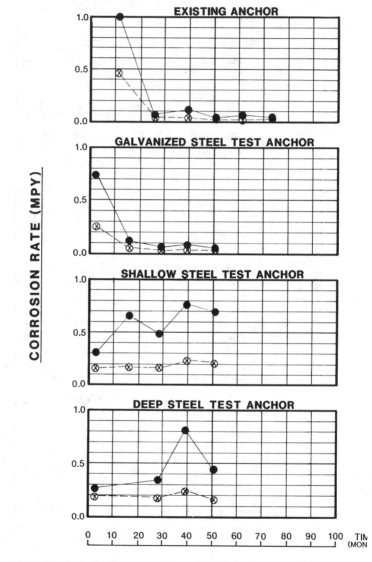

FIG. 11—*Corrosion rate versus time polarization data—Tower 303.*

TABLE 4—*Average thickness loss of test anchors after 4.3 years, mils.*

	Weight Loss	**ZINC** Accudern Zinc Thickness	Total Section Micrometer Thickness	Polarization Measurements PB	LP
TOWER 271					
GS-extension (8)	0.4	1.4	0		
GS-lead (26)	0.6	1.4	3		
Weighted average	0.5	1.4	1.2	0.5	0.9
TOWER 286					
GS-extension (2)	0.6	0.7	0		
GS-extension (1)	0.6	1.2	0		
GS-extension (3)	0.6	1.6	1		
GS-lead (25)	0.6	2.8	3		
Weighted average	0.6	1.7	1.3	0.6	1.3
TOWER 303					
GS-extension (7)	0.4	1.3	0		
GS-lead (27)	0.7	1.6	3		
Weighted average	0.5	1.4	1.5	0.7	1.8

	STEEL Tower 271 DS	SS	Tower 286 DS	SS	Tower 303 DS	SS
Weight loss	0.9	0.8	0.8	0.8	1.0	1.0
Micrometer	1 to 2	1 to 2	1	1 to 2	2	1 to 2
Polarization						
PB	1.1	1.1	2.1	2.1	1.7	1.6
LP	1.4	1.8	3.6	3.7	3.4	4.0
Corrosometer probes	0.3		1.1		0.3	

NOTE 1: For weight loss and polarization measurements, the assumed loss on the inside surface is equal to the loss on the outside surface.

NOTE 2: Losses shown are in mils and relate to loss in total thickness of test anchor.

NOTE 3: PB = polarization break; LP = linear polarization; DS = deep steel; SS = shallow steel; GS = galvanized steel.

Acknowledgments

I wish to thank Ed Escalante of the National Bureau of Standards for providing technical assistance in performing polarization testing. I would also like to thank Robert Grosshans of the Northern States Power Co. for funding the study.

References

[1] Romanoff, M., "Underground Corrosion," Circular 579, National Bureau of Standards, Washington, DC, April 1957.

[2] Romanoff, M., "Corrosion of Steel Pilings in Soils," Monograph 58, National Bureau of Standards, Washington, DC, October 1952.

[3] Schwerdtfeger, W. J., "Polarization Measurement as Related to Corrosion of Underground Steel Pilings," *Journal of Research of the National Bureau of Standards,* April 1971, pp. 107–121.

[4] Schwerdtfeger, W. J. and Romanoff, M., "Corrosion Rates on Underground Steel Test Piles at Turcot Yard, Montreal, Canada—Part I," Monograph 128, National Bureau of Standards, Washington, DC, July 1972.

[5] Schwerdtfeger, W. J. and McDorman, O. N., "Measurement of the Corrosion Rate of a Metal from Its Polarizing Characteristics," National Bureau of Standards, Washington, DC, 17 Mar. 1952, pp. 407–413.

[6] Jones, D. A., "Polarization Methods for Measuring the Corrosion of Metals Buried Underground," *Journal of Materials,* September 1969, pp. 600–617.

[7] Schwerdtfeger, W. J., "A Study by Polarization Techniques of the Corrosion Rates of Aluminum and Steel Underground for Sixteen Months," *Journal of Research,* National Bureau of Standards, October–December 1961, pp. 271–276.

[8] Stern, M. and Geary, A. L., "Electrochemical Polarization—I. A Theoretical Analysis of the Shape of Polarization Curves," *Journal of the Electrochemical Society,* January 1957, pp. 56–63.

[9] Mansfeld, F., "Tafel Slopes and Corrosion Rates from Polarization Resistance Measurements," *Corrosion,* National Association of Corrosion Engineers, October 1973, pp. 397–402.

[10] Compton, K. G., "Corrosion of Buried Pipes and Cables, Techniques of Study, Survey, and Mitigation," *Underground Corrosion, ASTM STP 741,* E. Escalante, Ed., American Society for Testing and Materials, Philadelphia, 1981, pp. 141-155.

[11] Jones, D. A., "Principles of Measurement and Prevention of Buried Metal Corrosion by Electrochemical Polarization," *Underground Corrosion, ASTM STP 741,* E. Escalante, Ed., American Society for Testing and Materials, Philadelphia, 1981, pp. 123-132.

[12] Stern, M., "Method for Determining Corrosion Rates from Linear Polarization Data," *Corrosion,* Vol. 14, No. 9, September 1985, pp. 440t–440t.

[13] Skold, R. V. and Larson, T. E., "Measurement at the Instantaneous Corrosion Rate by Means of Polarization Data," *Corrosion,* Vol. 13, No. 2, February 1957, pp. 139t-142t.

Edward Escalante[1]

Concepts of Underground Corrosion

REFERENCE: Escalante, E., **"Concepts of Underground Corrosion,"** *Effects of Soil Characteristics on Corrosion, ASTM STP 1013,* V. Chaker and J. D. Palmer, Eds., American Society for Testing and Materials, Philadelphia, 1989, pp. 81–94.

ABSTRACT: Corrosion in soil is a complex phenomenon, but there are some basic concepts that are useful in understanding the process. Underground corrosion is electrochemical in character, and this fact is used to describe the corrosion process in terms of an ordinary dry cell. The differences between corrosion in disturbed and undisturbed soil are discussed, and data are presented to emphasize these differences.

The results reveal that soil composition is less important than soil resistivity, but both are subordinate in importance to oxygen availability. Thus, corrosion is negligible in undisturbed soils where oxygen concentration is low.

KEY WORDS: corrosion in soil, corrosion of steel, disturbed soil, underground corrosion, undisturbed soil

Underground corrosion is the deterioration of metals or other materials brought about by the chemical, mechanical, and biological action of the soil environment. It is a process that can occur rapidly or very slowly, as will be described.

This paper presents an introduction to the concepts of underground corrosion and the factors that influence this corrosion in disturbed and undisturbed soils. Emphasis will be placed on relating these concepts to underground corrosion field studies.

Basic Concepts

Underground corrosion is electrochemical in character, a fact that allows us to examine the corrosion process by electrical means. Furthermore, the underground corrosion process is very similar to the electrochemical action that takes place in an ordinary dry cell of a flashlight during use. This type of dry cell is also called a galvanic cell.

A galvanic cell must have three components for it to function. These are: (1) an anode/cathode system; (2) an electrically conducting path between the anode and the cathode; and (3) an electrolyte in contact with the anode/cathode system. In the dry cell illustrated in Fig. 1, the zinc case and the carbon rod make up the anode/cathode system. The electrolyte is the chemical medium, normally an aqueous gel, between the zinc case and the carbon rod. The conducting path between the anode and the cathode is provided externally by the flashlight body which passes the current through the bulb for illumination.

In this type of dry cell, the zinc case is the anode which goes into solution in the electrolyte and thus corrodes in the process of giving up electrons for the production of electricity. This dissolution at the anode is referred to as an oxidation reaction. At this electrode a zinc atom gives up two electrons and becomes a positive ion. The electrons flow toward the cathode through the conducting metallic path while the positively charged ions either chemically combine with some other species or diffuse through the electrolyte toward the cathode where they

[1]Metallurgist, Corrosion Group, National Bureau of Standards, Gaithersburg, MD 20899.

DRY CELL

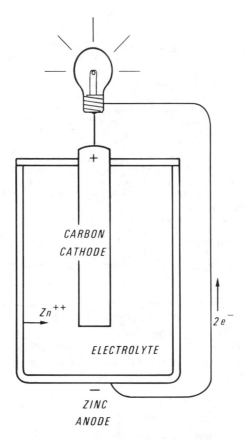

FIG. 1—*Galvanic cell (dry cell)*.

gain electrons and are reduced. Thus, the current is directly related to the dissociation of zinc, the corrosion process taking place at the anode.

The carbon rod in the dry cell is the cathode where reduction reactions occur. Positive ions are attracted to this electrode, where they are reduced by picking up one or more electrons at its surface. Reduced ions, such as hydrogen, can adhere to the cathode surface and form a barrier and diminish further reaction or they may diffuse away from the surface and allow the reactions to proceed. Since a reduction process rather than a dissolution process is taking place at the cathode, this electrode does not go into solution and is described as being under cathodic protection. The driving force for any galvanic cell is the potential difference developed between the anode and the cathode. This voltage difference can arise from a variety of conditions as follows.

One way in which a voltage differential develops is the bringing of two unlike metals into electrical contact. The difference in potential developed between the two metals and their relative chemical performance (anode or cathode) can be judged by examining a galvanic series as that shown in Fig. 2. This series is based on empirical results determined in seawater [1]. The materials are ranked from the most active (anodic) at the top to the most noble (cathodic) at the

GALVANIC SERIES IN SEAWATER

FIG. 2—*Galvanic series in seawater.*

bottom. Thus, if any two of these materials are connected in a galvanic cell as shown in Fig. 3, the more active material will act as an anode and corrode while the more noble material will be the cathode and be protected. For example, if we connect zinc and mild steel in seawater, the zinc, being higher on the series, will corrode (anode) while the steel is protected (cathode). On the contrary, if we connect mild steel to copper in seawater, the galvanic series shows that the steel will be anodic while the copper, being more noble, is cathodically protected.

Just as an anode/cathode system can develop between two unlike metals, a potential difference can develop on the surface of a single metal. Thus, differences in grain orientations, as illustrated in Fig. 4, can cause some grains to act as anodes while others act as cathodes with excellent electrical continuity existing in the bulk material. In addition, chemical anisotropy, inclusions, strained and unstrained areas, and other imperfections on the surface of a metal can give rise to these potential differences, providing a driving force for the corrosion process.

Inhomogeneities in the electrolyte, as illustrated in Fig. 5, can also cause potential differentials on a metal surface. Some examples of this are differential aeration, temperature differences, local depletion or accumulation of chemical species, or dissimilar rates of flow of the electrolyte. Oxides, precipitates, or other debris on the surface of the steel can reduce the amount of oxygen available locally and accelerate corrosion at that point. Similarly, crevices can affect availability of oxygen and cause localized corrosion. As previously mentioned, if any of the three components of a galvanic cell is removed, the electrochemical process stops. The

BATTERY OR GALVANIC CELL
(UNLIKE METALS)

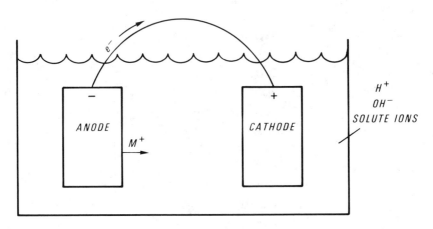

FIG. 3—*Galvanic cell (wet cell).*

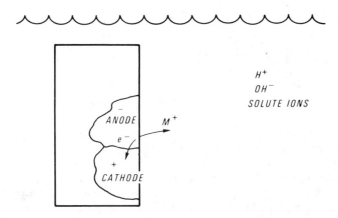

FIG. 4—*Potential differences due to metal surface anisotropy.*

driving force for the reactions are the potential differences that are generated on a metal or differences between two unlike metals. The effect of soil, the electrolyte, on the subsequent corrosion of metal will be the subject of the remainder of this paper.

Corrosion in Soil

The step from a relatively simple system such as a dry cell to the complex corrosion cell developed in an underground structure is a large one. However, the basic concepts of the corrosion process remain the same. But first, what is soil? Soil is defined by the Department of Agriculture as the loose surface material on the earth consisting of disintegrated rock with an admix-

DIFFERENTIAL ENVIRONMENT CELL

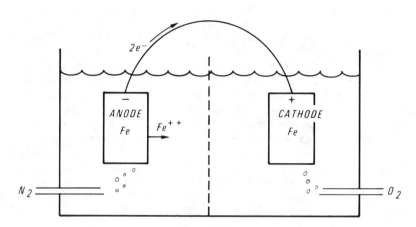

FIG. 5—*Potential differences due to differential aeration.*

ture of organic material on which plants grow. The interaction of these organic and inorganic materials gives rise to many variations in soil characteristics. Some of these characteristics have an important influence on underground corrosion as will be described later.

In general, the corrosion behavior of structural steel members in soil can be divided into two categories, that is, corrosion in disturbed soil and corrosion in undisturbed soil. A disturbed soil is a soil in which digging, backfilling, or other such soil upheaval has taken place. An undisturbed soil has not been disrupted in such a manner. Thus, driving a steel pile into the ground causes minimum change to the soil and is, relatively speaking, considered to be undisturbed. The effects on corrosion of these two conditions will be discussed separately.

Corrosion in Disturbed Soil

For studies which include the evaluation of soil corrosivity, the National Bureau of Standards (NBS) has maintained seven underground corrosion test sites in the United States. Over a period of years, a variety of specimens have been buried at each of these sites, and the soil corrosivity of each of these sites is now well known [2]. In addition, several factors that can influence the corrosivity of the soil have been evaluated over a period of time. Some of these factors are soil texture, internal drainage, resistivity, temperature, pH, and redox potential.

Soil texture is determined by the proportions of sand, silt, and clay that make up a soil. Clay, having the finest particle size and minimum pore volume between particles, tends to reduce the movement of air and water and can develop conditions of poor aeration when wet. Sand has the largest particle size and promotes increased aeration and the distribution of moisture. Soil texture, thus, has an important influence on the diffusivity of soluble salts and gases. It is worth noting that ASTM D 2487, Classification Of Soils For Engineering Purposes, describes a classification system based on particle size.

Internal drainage is that property of soil that describes the water retention properties of a soil and is related to soil texture. Internal drainage is also affected by the height of the water table.

Thus, a sandy soil which would normally have good permeation to moisture is considered to have poor internal drainage if the water table is high and keeps the soil in a saturated condition.

Soil resistivity is a measure of how easily a soil will allow an electric current to flow through it. This is also a measure of how effective the soil is as an electrolyte. The lower the resistivity of a soil, the better it will behave as an electrolyte and the more likely it is to promote corrosion. The measurement is described in the ASTM Standard Method for Field Measurement of Soil Resistivity Using the Wenner Four-Electrode Method (G 57-84). Table 1 shows a relationship between soil resistivity and soil corrosivity [3].

The temperature of the soil is certainly a factor in the corrosion process, and some interesting effects have been observed. For example, the resistivity of soil is inversely proportional to temperature, and an increase in soil temperature would be expected to increase the rate of the corrosion reaction. However, an increase in temperature also reduces the solubility of oxygen, which tends to reduce the rate of reaction at the cathode. The net result is that temperature does not have as large an effect on underground corrosion as one might expect [4].

Soil pH is the acidity or alkalinity of the soil media. Most soils and all loams are fairly well buffered, resulting in a soil pH that is not affected by rainfall. Sand, because of its high moisture diffusivity, can have its soluble salts leached out or diluted to the point that its pH will change during a heavy rain. Such changes appear to have little effect on corrosion of steel. An in situ measurement technique is described in the ASTM Test Method for pH of Soil For Use in Corrosion Testing (G 51-77).

The redox potential or oxidation-reduction potential is the potential of a platinum electrode in an electrolyte versus a reference half-cell converted to the standard hydrogen scale. The redox potential of a soil gives an indication of the proportions of oxidized and reduced species in that soil [5]. Very high corrosion rates have been observed in poorly aerated (reducing) soils where anaerobic bacteria often thrive. Starkey and Wight have studied soil redox potentials and have observed a relationship between redox potential and soil corrosivity as shown in Table 2 [6].

The location and soil properties of the seven sites are tabulated in Table 3. The column on the

TABLE 1—*Relationship between soil resistivity and soil corrosivity* [3].

Soil Resistivity, ohm-cm	Classification of Soil Corrosiveness
0 to 900	Very severe corrosion
900 to 2 300	Severely corrosive
2 300 to 5 000	Moderately corrosive
5 000 to 10 000	Mildly corrosive
10 000 to > 10 000	Very mildly corrosive

TABLE 2—*Relationship between redox potential and soil corrosivity* [6].

Range of Soil Redox Potential	Classification of Corrosiveness
Below 100 mV	Severe
100 to 200 mV	Moderate
200 to 400 mV	Slight
Above 400 mV	Noncorrosive

TABLE 3—*Properties of soils at test sites.*

Site	Soil Description	Location	Internal Drainage	pH	Redox Potential, V	Soil Resistivity, kohm-cm
A	Sagemoor sandy loam	Toppenish, WA	Good	8.8	...	6 to 26
B	Hagerstown loam	Loch Raven, MD	Good	5.4	+0.381	13 to 37
C	Clay	Cape May, NJ	Poor	4.3	+0.215	0.4 to 1.1
D	Lakewood sand	Wildwood, NJ	Good	5.6	+0.427	14 to 57
E	Coastal sand	Wildwood, NJ	Poor	7.1	+0.556	13 to 49
G	Tidal marsh	Patuxent, MD	Poor	5.6	−0.086	0.4 to 15
N	Glenville silt-loam	Gaithersburg, MD	Moderate	6.8	−0.273	7 to 10

left identifies each site by a letter arbitrarily chosen for convenience in referencing. The soil nomenclature in Column 2 is based on descriptions taken from soil survey maps published by the U.S. Department of Agriculture. Sites A through G were chosen as representative of corrosive and noncorrosive soils found throughout the United States. Availability and accessibility, of course, are of utmost importance. Most sites are located on government-owned property. Site N, located on NBS grounds, is used for short-term corrosion tests or testing of equipment. Each of the soil sites will be described on the basis of soil observations and measurements which will, in turn, be followed by a description of burial data results.

Site A—This testing site is located in the state of Washington outside of the Yakima Valley and is impressively surrounded by distant snow-covered mountains of volcanic origin. The region is semiarid with a rainfall of less than 25 cm/year (10 in./year) but supports an abundant growth of sagebrush. The area has a gentle slope leading to a nearby stream which is fed from a spring located about 50 m (160 ft) from the site. Being a sandy loam in a dry area, the internal drainage is very good. The soil surface layers are dry and powdery and show little evidence of moisture even to a depth of 1 m (3 ft). The soil pH is 8.8 and is the most alkaline of the seven sites. The redox potential was not measured at this site, but it is assumed that the soil is well aerated judging from its dry sandy texture. The resistivity of the as-received soil measured in the laboratory with a soil cup ranged from 6000 to over 26 000 ohm-cm. The average soil resistivity was over 10 000 ohm-cm but can be as low as 400 ohm-cm when wet. Long line metallic structures passing through these vastly different soil moisture conditions would be expected to develop corrosion problems. However, a structure confined to the high-resistivity soil of this arid region would not be seriously affected by corrosion.

Site B—Loch Raven, Maryland, is located north of Baltimore and within 2 km (1.2 miles) of a water reservoir. The area is gently rolling, densely wooded hills with a natural cover of weeds and grasses. Rainfall is of the order of 125 cm/year (50 in./year). The soil has good drainage and is made up of a loam to a depth of about 30 cm (12 in.) which is underlain by a red-brown clay. The pH of the soil is 5.4 and has a redox potential averaging +381 mV, indicating good aeration. The earth resistivity is high and ranges from 13 000 to 37 000 ohm-cm with an average of about 33 000 ohm-cm. The relatively low pH and the clay soil beneath led to corrosive conditions in the area of the waterline where oxygen concentration cells develop. The high resistivity, however, tends to reduce this effect, resulting in generally low corrosion.

Site C—Cape May, New Jersey, is located at the mouth of Delaware Bay facing the Atlantic Ocean. The area is generally flat and devoid of hills. Site C, which is a large clay/silt pit, is located about 100 m (320 ft) from the ocean. Permeability of this clay/silt is so low that rainfall runs off of its surface with little percolation. The pH of this soil is acid, having a value of 4.3.

The redox potential of this soil fluctuates but tends to be positive at about $+215$ mV, suggesting a moderately corrosive environment. The earth resistivity is consistently below 1000 ohm-cm with an average of about 600 ohm-cm. The close proximity of the ocean combined with the very low resistivity of the soil makes this a very corrosive site.

Sites D and E—The conditions at these sites will be described together because they are very similar in most respects except that Site E is a coastal sand located on the beach nearest the ocean, while Site D is a few hundred metres inland. These two sites are along the Atlantic Ocean approximately 1 km (0.6 miles) north of Cape May. Though both are loose sand soils, Site E nearest the ocean is often saturated with saltwater with poor drainage, while the inland site is rarely, if ever, flooded. The inland site supports an abundant growth of shrubs and grasses, which periodically must be cleared for access to the site. The beach site is beginning to develop a growth of beach grass, which builds and holds dunes in the area. The specimens that were originally 1 m (3 ft) below the surface are now 2 to 3 m (6 to 9 ft) underground because of the growth of the sand dunes. The soil pH of the inland site is 5.6, while the beach site is essentially neutral at 7.1. Both exhibit high redox potentials well over $+400$ mV, indicating good aeration. The earth resistivities at both of these sites fluctuate with time. Their high permeabilities allow soluble salts to wash away during a heavy rainfall, resulting in unusually high resistivities after such a rain. These fluctuations are greatest for the beach sand, which has ranged from 13 000 to 49 000 ohm-cm. The average resistivity for the inland site (Site D) is 33 000 ohm-cm, while the beach site's (Site E's) average resistivity is 24 000 ohm-cm. The uniformity in the makeup of this soil and the high resistivities tend to make these low-corrosivity soils.

Site G—Patuxent is situated near the southern tip of Maryland and lies alongside the Chesapeake Bay. The area is hilly and heavily wooded with a luxuriant growth of weeds and grasses. The test site is in a creek estuary, which is commonly flooded during high tide. Its internal drainage is very poor because of the clay type of the soil, which holds water, and because of the water table, which is very close to the surface. The pH of this soil is maintained at a pH of about 6 by sulfate-reducing bacteria, but if a soil sample is removed and allowed to stand for at least 48 h, its pH drops to less than 4.0. As the sulfide-rich soil is exposed to oxygen, the sulfide oxidizes to sulfate, causing the pH to drop [7]. This is an extreme example of the importance of making pH measurements in situ rather than on samples carried back to the laboratory. The redox potential of this soil is negative, indicating poor aeration. Its resistivity is often below 500 ohm-cm but can fluctuate to over 15 000 ohm-cm. During high tide, salt water flows upstream, often flooding the site and lowering its resistivity. At low tide, however, the site is again exposed to fresh water, which, when combined with heavy rainfall, raises the resistivity of the soil. The combined effects of these changing conditions and the presence of anaerobic bacteria make this soil very corrosive.

Site N—This site is located within the grounds of NBS in an area surrounded by woods. Rainfall in this region, as in most of Maryland, is of the order of 125 cm/year (50 in./year). The land is gently sloping and easily supports the growth of grass. The top soil is about 30 cm (1 ft) with a siltpan subsoil with moderate to poor drainage. This siltpan is plastic and clay-like when wet. The pH of this soil is 6.8 and is very well buffered. Site B is similar to Site N in this respect. The redox potential of the NBS site is consistently negative, indicating poor aeration at a depth of 1 m (3 ft), which is well within the silty subsoil. Its resistivity ranges from 7000 to 10 500 ohm-cm with an average of 8500 ohm-cm. Like most loams, the resistivity of this soil is reduced by rainfall, which is opposite to the effect observed in sand soil. In spite of its low redox potential, this site is not corrosive because of poor moisture diffusion and high earth resistance. However, long line structures passing through this poorly aerated soil from a well aerated soil could suffer corrosion damage.

Observed Corrosion—The corrosiveness of the seven sites is demonstrated by the results of the following two studies. Figure 6 is a plot of the weight loss of carbon steel at the seven sites over an eight-year period [8]. Illustrated in Fig. 7 is the average galvanic current generated

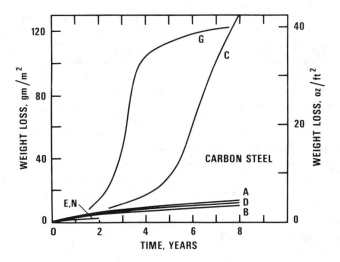

FIG. 6—*The underground corrosion of steel versus time at seven test sites.*

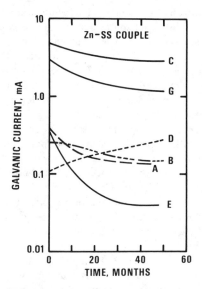

FIG. 7—*The underground corrosion of zinc galvanically connected to stainless steel versus time at six underground test sites.*

between a stainless steel specimen and a zinc specimen of equal area [155 cm² (24 in.²) with zinc serving as a sacrificial anode] [4]. Both corrosion studies clearly indicate that the soils at Sites C and G are an order of magnitude more aggressive than the soils at the other five sites.

Attempts to relate the various measured soil parameters with the observed corrosion were only partially successful. On the basis of Table 2, for example, the redox potential indicated that the soil at Site N is corrosive and that the soil at Site C is slightly corrosive. The corrosion data in Figs. 6 and 7 demonstrate that this indication is not correct and reveal that Site C is very corrosive compared to Site N. In fact, the redox potential of the soil was found to be a poor indicator of the soil aggressiveness. It is likely that this difficult redox potential measurement is not sufficiently reliable for determining the reduction potential of soil. Similarly, a poor relationship was found to exist between soil pH and corrosion. For example, the two most acid soils, Sites B and C, are on opposite extremes of soil corrosivity as indicated by the observed corrosion. The only measurement that did show any relationship to the soil aggressiveness was the resistivity of the soil. Sites C and G, the most corrosive soils, on several occasions had resistivities below 500 ohm-cm. As indicated in Fig. 6, the soils at Sites A, B, D, E, and N were similar in their rate of attack towards steel. The highest resistivities were measured at Sites B and D, and on this basis might be expected to produce the lowest corrosion rates. However, the corrosion rates at Sites B and D are higher than at Site E, which has a lower resistivity. In general, disturbed soils with low resistivities (500 ohm-cm or below) are indicative of highly corrosive conditions. However, as the resistivities increase above 2000 ohm-cm, the soil corrosivity-resistivity relationship becomes less reliable [9].

Corrosion in Undisturbed Soil

Early studies on steel piles driven underground indicated that the corrosion observed over a period of several years was much less than expected on the basis of disturbed soil data, and, furthermore, it was noted that this corrosion was independent of the soil conditions [10].

The role of oxygen in an undisturbed soil overrides the effects of soil resistivity, pH, etc. described in the previous discussion on disturbed soil where oxygen is more readily available. The diffusion of oxygen in undisturbed soil, and particularly below the water line, is sufficiently low that the corrosion process is effectively stifled. In those situations where a steel pile is driven into undisturbed soil and then backfilled with a soil layer, the section of the pile in the disturbed soil is cathodic to the rest of the pile in the undisturbed region as illustrated in Fig. 8. As a result, the most severe corrosion occurs on the section of the pile just below the disturbed layer. The effectiveness of this anode/cathode system is dependent on the oxygen differential between the disturbed and undisturbed layers, which gives rise to potential differences between these zones. The resistivity of the soil determines the distance over which the anode/cathode extends. With a low-resistivity soil, the surface area of the anode in the disturbed layer is larger and the attack is more general in nature. With a high-resistivity soil, the corrosion is confined to a smaller area, resulting in localized corrosion. Within the disturbed soil region, differences in potential due to soil anisotropy can develop on the surface of the steel. These local short range cells can override the cathodic tendency of this section of the pile and develop significant localized corrosion attack.

Similarly, a pile located in undisturbed soil with a high water table can suffer some corrosion attack at the waterline as illustrated in Fig. 9. This combination does not result in serious attack, but it is believed that the situation is aggravated by a continuously changing water table, which would draw in oxygen as the waterline dropped. In this case, as in the one before, the section of the pile above the waterline acts as a weak cathode to the anode below the waterline.

Steel piles are commonly concrete capped when used as supports for large structures. The area of the steel in the concrete forms a passive oxide film generated by the action of the highly alkaline concrete environment, and this area is cathodic to the rest of the pile in the soil. Fortunately, the high resistivity of the concrete limits the effectiveness of the cathode, and the small amount of corrosion attack that results is concentrated in the region of the pile immediately outside of the concrete as illustrated in Fig. 10. Initially, the corrosion rate is high while oxygen is available. Once the oxygen is depleted, the corrosion rate drops and is controlled by the slow

DIFFERENTIAL ENVIRONMENT CELL

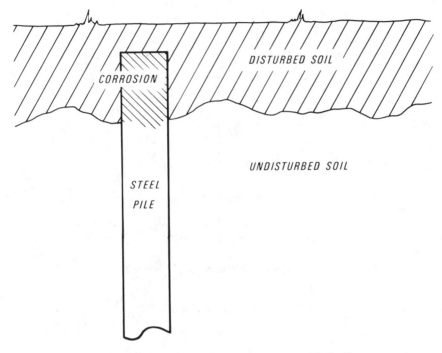

FIG. 8—*The corrosion of a steel pile in a disturbed soil.*

diffusion of oxygen through the soil and concrete. Figure 11 illustrates some results on a concrete capped steel pile (C-4) after seven years exposure. Briefly, the soil conditions at this site are as follows. The average resistivity in the soil in the first 3 m (10 ft) was less than 1500 ohm-cm, and below this the resistivity averaged 3900 ohm-cm. The average pH for the soil was 7.4. The top surface of the soil is a fill made up of cinders, gravel, sand, and some clay. The data show that the maximum metal loss occurred just below the concrete and is of the order of 0.07 mm/year (2.8 mils/year), and, over the remainder of the pile, the loss averaged 0.02 mm/year (0.70 mils/year) over a seven-year period. Thus, even where the attack was concentrated below the concrete, 95% of the pile is intact, while the remainder of the pile is 99% intact. It was estimated that between 50 to 75% of the mill scale remained over most of the pile except in a 2-m (6-ft) section below the concrete.

Though the corrosion observed on the driven piles is not extensive, there are ways in which it can be reduced even further. One direct approach that would minimize corrosion on a driven pile in all three cases discussed is to reduce the effectiveness of the cathode by coating the section of the pile above the waterline in the disturbed soil and, in particular, the area of the pile in the concrete. Thus, by removing the cathode, the anode/cathode system is disrupted, resulting in reduced corrosion.

DIFFERENTIAL ENVIRONMENT CELL

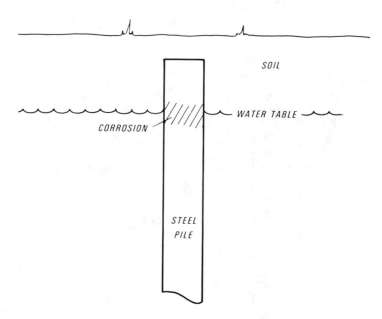

FIG. 9—*The underground corrosion of a steel pile at the waterline.*

DIFFERENTIAL ENVIRONMENT CELL

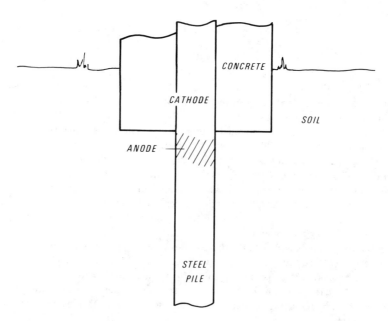

FIG. 10—*The underground corrosion of a steel pile with a concrete cap.*

CORROSION OF CONCRETE CAPPED STEEL PILE (C-4)
7 YEARS EXPOSURE

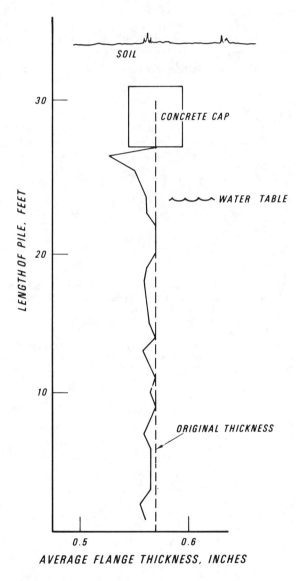

FIG. 11—*A plot of flange thickness along the length of a pile with a concrete cap after seven years exposure.*

Summary

Underground corrosion is an electrochemical process requiring the existence of three conditions. To be effective the corroding system must have: (1) an anode/cathode system; (2) an electrically conducting path between the anode and the cathode; and (3) an electrolyte in contact with the anode cathode system. Remove any of these and the corrosion process stops.

Corrosion in disturbed soil is a function of the soil environment, but soil pH and redox potential are poor indicators of a corrosive soil. A soil with a resistivity below 500 ohm-cm is corrosive. Above 2000 ohm-cm the relation of soil resistivity to soil corrosivity is less reliable.

The corrosion of steel piles in undisturbed soil is independent of the soil environment. Even with low soil resistivities, the corrosion observed is very low. Coating the cathodic area of the pile in the disturbed soil zone above the waterline or in the concrete cap will further reduce corrosion effects.

References

[1] Laque, F. L., *Marine Corrosion Causes and Prevention*, John Wiley and Sons, New York, 1975.
[2] Escalante, E., "Corrosion Testing in Soil," *Metals Handbook*, Vol. 13, American Society for Metals, Metals Park, OH, September 1987.
[3] Waters, F. O., "Soil Resistivity Measurements for Corrosion Control," *Corrosion*, Vol. 8, No. 12, 1952, p. 407.
[4] Escalante, E., "The Effect of Soil Resistivity and Soil Temperature on the Corrosion of Galvanically Coupled Metals," ASTM Symposium on Galvanic Corrosion, Phoenix, AZ, November 1986, in *Galvanic Corrosion, ASTM STP 978*, in press.
[5] Iverson, W. P., "Tests in Soils," *Handbook on Corrosion Testing and Evaluation*, W. B. Ailor, Ed., John Wiley and Sons, New York, 1971, p. 575.
[6] Starkey, R. L. and Wight, K. M., "Anaerobic Corrosion of Iron in Soil," American Gas Association, New York, 1945, p. 99.
[7] Iverson, W. P., "An Overview of the Anaerobic Corrosion of Underground Metallic Structures, Evidence for a New Mechanism," in *Underground Corrosion, ASTM STP 741*, American Society for Testing and Materials, Philadelphia, 1981, p. 33.
[8] Romanoff, M., "Performance of Ductile Pipe in Soils," *Journal of American Water Works Association*, Vol. 60, No. 6, June 1968.
[9] Shepard, E. R., "Pipe Line Currents and Soil Resistivity as Indicators of Local Corrosive Soil Areas," *Bureau of Standards Journal of Research*, Vol. 6, No. 4, April 1931.
[10] Romanoff, M., "Corrosion of Steel Piles in Soils," NBS Monograph 58, National Bureau of Standards, Gaithersburg, MD, 1962.

Richard A. Corbett[1] and Charles F. Jenkins[2]

Soil Characteristics as Criteria for Cathodic Protection of a Nuclear Fuel Production Facility

REFERENCE: Corbett, R. A. and Jenkins, C. F., **"Soil Characteristics as Criteria for Cathodic Protection of a Nuclear Fuel Production Facility,"** *Effects of Soil Characteristics on Corrosion, ASTM STP 1013,* V. Chaker and J. D. Palmer, Eds., American Society for Testing and Materials, Philadelphia, 1989, pp. 95–106.

ABSTRACT: The fact that buried metallic structures corrode is well documented. It has been postulated that the extent and rate of attack is controlled predominantly by the characteristics of the surrounding soil. Therefore, prior to constructing a new facility designed to process accumulated nuclear waste, consideration was given to protecting its underground pipelines against corrosion. Leak frequency curves from other nearby plant sites, extensive soil resistivity surveys, and geochemical analysis were used to evaluate the on-site soil characteristics for corrosion susceptibility.

Analysis of the data collected over a three-year period indicated that although the soil is not overly aggressive, substantial heterogeneity existed so as to establish galvanic cells along pipe lengths passing through the soil. To limit the extent of corrosion on underground piping, the application of an impressed current cathodic protection system was recommended to supplement a high-integrity, corrosion-resistant coating and wrap system.

KEY WORDS: cathodic protection, corrosion, soil characteristics, soil surveys, nuclear, underground, site selection

Buried metallic structures will corrode as a result of electrochemical activity of the soil. Therefore, in 1980, prior to constructing a new facility designed to process accumulated nuclear waste sludge at the Savannah River Plant (SRP), extensive soil resistivity surveys and geochemical analyses were performed to evaluate the aggressiveness of the existing soil and its specific characteristics.

The Defense Waste Processing Facility (DWPF) is designated to receive radioactive wastes from SRP nuclear fuel production in a liquid slurry form and encapsulate it into a permanent solid glass form. The wastes from the chemical separations process and tank farm storage areas will be transferred through underground piping systems for up to five miles. Because of the radioactive nature of the slurry, special care utilizing conservative design and installation approaches are applied throughout. Public safety demands assurance that no failures occur during the reasonable design life of the entire system. This assurance requires apparent redundancy in several modes of protection (that is, coating, wrapping, special backfill, cathodic protection, etc.).

The radioactive wastes have been stored temporarily in liquid form since startup of SRP in 1953. Currently, 30 million gallons exist in 51 storage tanks, and additional new waste is accumulating at the rate of one million gallons per year. It is necessary to provide permanent storage

[1]Corrosion Testing Laboratories, Inc., 410 B & O Lane, Wilmington, DE 19804.
[2]E. I. duPont de Nemours and Co., Savannah River Laboratory, Aiken, SC 29808.

to alleviate the need for additional temporary storage facilities and to allow for emptying and abandonment of the oldest tanks. Throughout SRP, buried pipelines used in transferring any form of radioactive waste consist of austenitic stainless steel core pipe within a carbon steel pipe jacket. Several different modes of corrosion protection have been used in the past for the outer jacket, including painting or coating, wrapping, surrounding, or encapsulating the jacket within hydrophobic insulating materials, etc. Backfill, aside from the insulation material, is typically a clay soil selected to slow migration of spilled or leaked species, should this occur. However, none of the protective techniques has been entirely satisfactory in avoiding corrosion. Cathodic protection has not yet been applied to any of the existing waste transfer lines.

Background of On-site Underground Corrosion

Backfill for buried piping and other equipment will generally be the excavated soil; but imported, lower permeability clay fill may also be used if the excavated soil is too porous. The soil represents the last controllable means of protection against contamination of the water table and nearest aquifer. If a leak should develop, it is desirable to maintain a low rate of seepage. Low-permeability, impervious clay provides the slowdown of percolation that is desired [1]. This is due to three characteristics of clay: absorption of water, swell, and ion exchange.

The swelling capability of clays upon wetting improves the already low permeability. However, there is a negative effect in the tendency of the wetted clay to hold moisture in the vicinity of buried lines, thus permitting aqueous corrosion of carbon steel. If the soil is high in soluble salts or if it has high total acidity and is alternately wet and dry, it may be especially corrosive [2].

To review the effects of corrosion on underground piping, a history of buried piping failures in the SRP waste farms was compiled for 1974 to 1986. Until now, nearly all failures of buried carbon steel piping have been on service piping, that is, steam, air, and water lines (Fig. 1). But corrosion penetrations of carbon steel jacket piping and of leak detection piping on process lines have been observed. Simultaneous leakage from the encased stainless steel core lines has not occurred, thus avoiding radioactive discharges to the soil. The failures of primary concern are those due to external soil side corrosion, and they are included on an accumulated leak frequency chart (Fig. 2).

The corrosion curve in Fig. 2 is not linear but rises at an increasing rate. A projected significant increase in the number of leaks and associated contingent problems exists for several years. The curve is classical in nature [3] and follows general experience with underground corrosion. This means that once leaks start to occur, an increase in their rate of development can be anticipated [4]. Analysis of Fig. 2 reveals that the total failure curve follows the same slope as the corrosion curve. This implies that corrosion is a major influence in the failure rate of underground piping at SRP. Better coatings and use of cathodic protection would provide for substantially fewer failures.

Also included in Fig. 2 is a total failures curve. This indicates occurrence of failures in addition to those caused by external corrosion. Included, for example, are breaks in cast iron lines used for clean well water. Such failures often follow in an area where previous excavation has occurred, for example, during repair of a leaking pipe. Disturbance of soil, disturbance of compaction, and use of heavy equipment all contribute to failures in cast iron piping and can sometimes be related to later corrosion occurrences in an area.

DWPF Site Selection

Several factors affected the choice of a location at SRP for construction of the DWPF facility.[3] These factors included a need to be near the existing tank farm areas to minimize transfer

[3]Internal report, E. I. du Pont de Nemours & Co., 1982.

FIG. 1—*Leak in steam condensate line in waste farm area. The line was under no corrosion protection. the adjacent steam line was buried in a hydrophobic backfill.*

distances for the radioactive wastes. The existing waste storage areas are near the center of the plant, adjacent to the principal separations facilities that generate the waste. The factors used for site selection are:

1. Maximize distance to plant boundaries.
2. Be positioned at relatively high elevations.

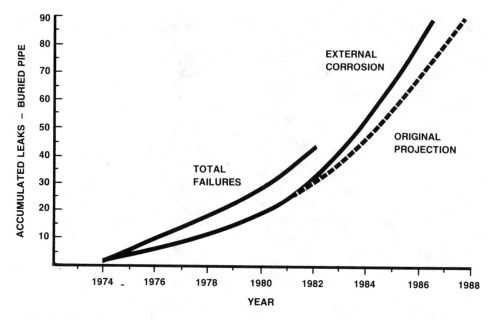

FIG. 2—*Accumulated leak data for underground pipelines in the SRP waste tank farms. Original projection was in 1982.*

3. Have few open flowing streams in order to reduce the risk of accidental release to such streams.

In addition, utilization of existing roads and railroads and employee travel constraints were also considered.

The DWPF facility is located in S Area, a gently rolling area adjacent to the H Area (Fig. 3). All high-level radioactive waste will transfer from or through H Area to the DWPF. F Area waste will first be delivered to H Area holding tanks and then to S Area.

The site elevation is relatively high at 250 to 300 ft above sea level (Fig. 4). Interarea transfer line piping will be kept above the water table. Groundwater drainage will be primarily to Upper Three Runs Creek, approximately 0.8-mile northwest of the S Area (Fig. 3). Southeasterly drainage to McQueen Branch will also end up in Upper Three Runs.

Kaolinite is the predominant clay in the soil, but montmorillonite was used in early construction to backfill some areas in the F and H Area waste farms. The basis for the choice of backfill material is in the relative ion-exchange capabilities of the two clays: montmorillonite is a finer particle and has a typical ion-exchange capacity seven to ten times that of kaolinite [4]. Ion exchange is a factor in controlling rate of vertical movement of spilled contaminants.

Clay is considered impervious. As they are wetted, clay aggregates become dispersed and permeability decreases. Dilute monovalent ions, such as Na^+, in the presence of divalent and trivalent cations such as Mg^{+2}, Fe^{+3}, and Al^{+3}, can prevent the soil from aggregating and permeability from increasing. Since the waste solutions contain very high concentrations of sodium and concentrations of other cations sufficient to prevent recombination as aggregates, any waste spill will be contained.

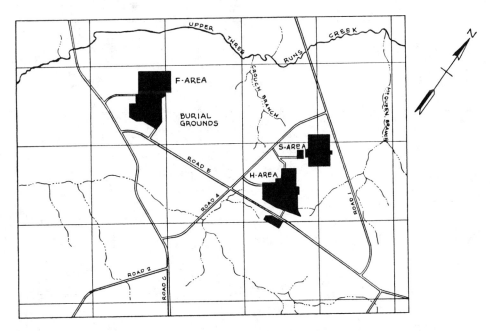

FIG. 3—*DWPF area plot plan: S Area.*

Standard On-Site Survey

The Need and Importance

Metallic corrosion of underground structures at the SRP site has been documented above. This corrosion is a physical and electrochemical process that requires soil moisture to form solutions with dissolved salts. Any factor that influences the soil solution also affects the corrosion cell. These factors include soil pH, resistivity (or specific conductance), concentration of chlorides, sulfates and oxygen, alkalinity, and potential acidity of the soil.

Materials engineers evaluate the corrosivity of soil on buried metallic structures based on:

1. Soil resistivity.
2. Total acidity.
3. Soil composition.
4. Soil drainage and water table.

The correspondence between the physical and chemical characteristics of soils and their corrosivity toward metals has been published in the National Bureau of Standards Circular 579 [5]. The most commonly agreed-upon criteria to rank the degree of corrosivity among soils is resistivity (Table 1) and total acidity, although many other variables should be considered to correlate soil characteristics with actual corrosion failures.

Review of Physiography and Geology

SRP is located in the South Carolina Atlantic Coastal Plain Province approximately 27 miles southeast of Augusta, Georgia. The Atlantic Coastal Plain Province is divided into the northeast trending Upper, Middle, and Lower Coastal Plain subprovinces [6].

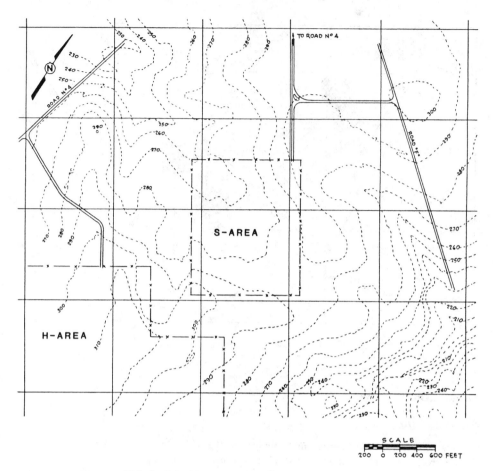

FIG. 4—*S Area topography: ground surface contours.*

TABLE 1—*Soil corrosivity versus resistivity.*

ohm-cm	Description
Below 500	Very corrosive
500 to 1000	Corrosive
1000 to 2000	Moderately corrosive
2000 to 10 000	Mildly corrosive
Above 10 000	Progressively less corrosive

The site is situated in the Upper Coastal Plain subprovince and is underlain by a sedimentary wedge approximately 1050 ft thick. This wedge consists of unconsolidated and semiconsolidated marine and alluvial sediments that dip gently to the southeast and range in age from Upper Cretaceous to recent. The sedimentary wedge is underlain by a Precambrian and Paleozoic basement complex composed of the folded and faulted igneous and metamorphic rocks of the

Piedmont physiographic province. Within the basement complex are structural basins containing unmetamorphosed volcanic and sedimentary nodes of probable Triassic-Jurassic age.

The surface soils at the DWPF site are from the Barnwell Formation, consisting of clays, sandy clays, and clayey sands. This is successively underlain by the McBean, Congaree, Ellenton, and Tuscaloosa formations (Fig. 5). The subgrade soils generally consist of silty and clayey sands to a depth of approximately 70 ft. All soils appear to be normally consolidated, though the McBean has a layer of calcareous clay with numerous small cavities. Sands vary in density from loose to medium. Clays and silts generally vary in consistency from medium stiff to stiff and have a low water permeability.

Stage One Investigation—Initial Soil Resistivity Survey and Geochemical Analysis

As part of the process for the DWPF site selection, an initial soil resistivity survey was performed in 1980. A total of 26 measurements were made at 20 ground locations around the proposed DWPF site in S Area (Fig. 4). The "four pin" method [7] was employed. The purpose of this initial survey was to define the soil resistivity characteristics of the subsurface with emphasis on determining the near-surface characteristics. This information would then be available for possible use in the design of ground beds for a cathodic protection system.

Results of the soil resistivity survey (Table 2) indicate occurrence of numerous lateral and vertical changes between sands and clays over short distances, which is typical of the coastal plain environment. As expected, localized zones of either highly conductive or highly resistant material were observed at or near the surface. In some cases, these zones appeared to be only 1 to 2 in. thick. The resistivity measurements indicated there were only three to five distinct layers over the site. However, the DWPF site may be generalized as having a variable and localized

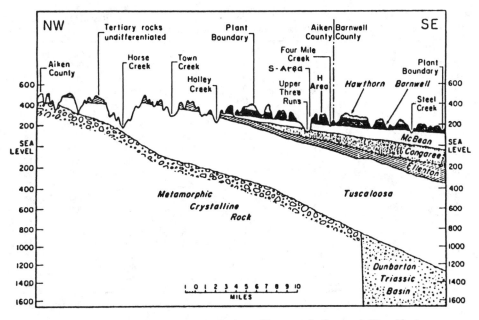

FIG. 5—*Generalized NW to SE geologic profile across the Savannah River Plant.*

TABLE 2—*Soil resistivity (ohm-cm) for DWPF site.*

Depth (feet)	Low	High	Average
0 to 5	3000	1 330 000	214 000
5 to 10	8000	308 000	78 000
>10	300	430 000	99 000

surface resistivity distribution. Beneath the surface is a resistant layer (120 000 to 300 000 ohm-cm), which varies in thickness from 15 to 40 feet and which overlies the groundwater table. The material below the groundwater table acts as a basement with an observed resistivity of 300 to 20 000 ohm-cm.

Soil samples from nine geotechnical borings were obtained for geochemical testing. The tests were performed to determine pH, specific conductance, chlorides, sulfates, and potential acidity of the subsurface soils (Table 3).

Stage Two Investigation—Soil Survey for Surrounding Areas

Results of soil surveys in the existing F and H Area waste farms have indicated a generally high resistivity (30 000 to 149 000 ohm-cm) with pockets of lower resistance down to 2000 to 8000 ohm-cm (Table 4). At the lower resistivities, the soil may be locally corrosive. Newer waste transfer lines are buried in a hydrophobic, inorganic powder insulation (>100 000 ohm-cm) and may be protected because of its very high resistivity.

TABLE 3—*ASTM Method A leachate quality.*[a]

Boring Number	Alkalinity[b] mg/L CaCO₃	Acidity[b] mg/L CaCO₃	Chloride, mg/L	Total Hardness[b] mg/L CaCO₃	pH	Resistivity, ohm-cm	Sulfate, mg/L
1	4	6	<0.5	<2	5.60	67 000	2
2	3	5	1.4	<2	6.05	125 000	6
2	3	4	1.4	<2	5.75	67 000	12
3	4	11	1.9	<2	5.40	24 000	20
3	3	10	1.0	140	5.20	36 000	14
3	3	8	1.0	<2	5.50	47 000	16
3	3	5	<0.5	<2	5.40	167 000	23
4	4	6	1.2	54	6.00	16 000	34
4	3	11	1.7	100	5.50	16 000	28
5	6	<2	1.2	<2	6.20	36 000	14
5	5	3	1.4	<2	6.15	20 000	30
5	6	3	1.7	<2	6.35	36 000	24
6	4	7	0.7	92	5.25	17 000	38
7	8	<2	<0.5	<2	6.30	83 000	69
7	4	12	2.1	<2	5.20	33 000	16
7	3	14	1.7	8	5.05	30 000	20
7	4	8	1.9	6	5.30	33 000	22
8	3	13	1.4	<2	5.05	28 000	20
9	3	3	1.7	<2	5.60	111 000	6
9	3	3	1.9	<2	5.70	100 000	6

[a]The leachate generated from ASTM Method A is representative of one part soil to four parts distilled water.

[b]mg/L = milligrams per liter ≃ parts per million.

TABLE 4—*Soil resistivity, waste tank farms.*

Depth (feet)	Wenner 4-Pin Resistivity, ohm-cm		
	Minimum	Maximum	Average
F AREA			
2.5	14 800	90 800	37 600
5.0	10 800	114 700	37 900
7.5	8 700	143 400	36 400
10.0	6 700	143 000	34 100
15.0	2 300	149 100	27 800
H AREA			
5	17 000	100 000	39 200
10	14 500	127 500	63 400
15	5 000	60 000	30 400
20	5 000	132 000	39 300

Large variations in soil resistivity provide for a possibility of galvanic couples. For example, a galvanic corrosion phenomenon can occur between parallel lines that are electrically connected (continuous). This occurs when the lines are buried in different media having different resistivities. Numerous lines in the waste farm areas fall into this category, where process and steam lines are buried in the hydrophobic insulating envelope and nearby lines are directly buried. Also, where leak detection piping crosses from the insulation envelope into adjacent ordinary clay or dirt fill of relatively low resistance, it is subject to potentially high anodic current flow at the interface due to the high potential difference across the interface.

"Long Line" corrosion (concentrated corrosion) on underground pipelines has also been experienced at SRP. This is a selective type of corrosion that attacks only portions of a given length of pipe. This type of corrosion usually appears when the pipe traverses soils of different composition and, as a result, one section of the pipe becomes anodic with respect to another.

Stage Three Investigation—Determine If Cathodic Protection Is Required

After review of the data generated from the initial survey and comparing it to that information available from the F and H Areas, an additional survey was performed to determine if cathodic protection was required to mitigate underground corrosion in the S Area. Soil resistivity measurements were taken in the S and H Areas along the proposed path of the interarea transfer line to confirm existing data. Table 5 shows the range of readings taken from the H Area to the S Area. The data indicate that the soil is moderately corrosive in the H Area and mildly corrosive in the S Area. A considerable difference in resistivities was again noticed from

TABLE 5—*Soil resistivity (ohm-cm) along interarea waste transfer line.*

Area	Low	High	Average
S	11 000	484 000	152 000
H	14 000	412 000	79 000

one location to another. These dissimilar soils, existing along the transfer piping right-of-way, would very likely cause accelerated galvanic corrosion in the low resistivity locations.

Seven soil samples were obtained along the proposed path and analyzed for pH, chlorides, sulfates, calcium, and magnesium (Table 6). All samples, except No. 6, showed high resistivity and low total dissolved solids. The pH of all samples was neutral. This indicates that the groundwater could be more corrosive than anticipated because protective calcareous deposits will not be able to form at cathodic sites on the metallic surfaces. A study of the soil samples with water of various saturation indices indicated a corrosive tendency of the soil towards leaching by groundwater.

Based on the data collected and compiled, and in concert with Du Pont Engineering Standards, the application of an impressed current cathodic protection system for the DWPF project was recommended to supplement coating, wrapping, and encapsulation in a hydrophobic backfill. This decision was based on a conservative approach including:

1. Heterogeneous soil resistivities that could lead to galvanic corrosion.
2. Soil chemistry leading to a corrosive tendency.
3. A leak frequency history in adjacent areas.

A closely coupled distributed and impressed current cathodic protection system was chosen for S Area. Such a system will limit the amount of current discharge per anode to reduce voltage gradients around the anodes. This system will also minimize the detrimental effects of stray currents occurring on discontinuous structures. The impressed current system can easily be adjusted to increase or decrease the current and voltage required for the protection of the underground structures.

Cathodic Protection Design for DWPF

Based upon the analysis of data obtained since 1980, an impressed current, closely distributed cathodic protection system was recommended to control underground corrosion of the piping system in the DWPF S Area.

The cathodic protection system will consist of six rectifiers and 223 high-silicon iron, prepackaged anodes (Fig. 6). The system is designed to protect all underground metallic structures for a minimum of 25 years. At locations where pipes are fitted together through bell and spigot joints (that is, fire water lines), electrical jumper bond wires are welded across the joints (Fig. 7).

TABLE 6—*Soil chemistry along interarea waste line.*

Sample Number	1	2	3	4	5	6	7
Calcium, mg/L	4.0	36	4.0	4.8	3.2	124	4.0
Magnesium, mg/L	1.5	3.4	1.0	1.9	1.0	2.4	1.5
Alkalinity, mg/L	6.1	39	8.5	12.2	14.6	97.6	8.5
Sulfate, mg/L	5.8	80	7.0	12.0	10.0	195	7.6
Chloride, mg/L	1.0	3.5	2.5	7.5	7.5	6.0	5.0
pH	7.15	7.18	7.17	7.05	7.18	7.20	7.20
Conductivity, μmho/cm	30	250	40	80	90	640	50
TDS, mg/L	20	150	26	52	50	410	32
Corrosivity, mils per year	0.6	1.0	1.4	0.7	0.5	0.5	1.1
Pitting index	0.5	1.0	4.0	0.1	0.1	0.1	0.8
Stability index	13.11	9.58	12.83	12.37	12.52	7.2	12.58
Saturation index	−2.98	−1.20	−2.83	−2.66	−2.67	−0.25	−2.69
% moisture of soil	18.2	10.1	10.9	10.6	11.5	11.4	8.8

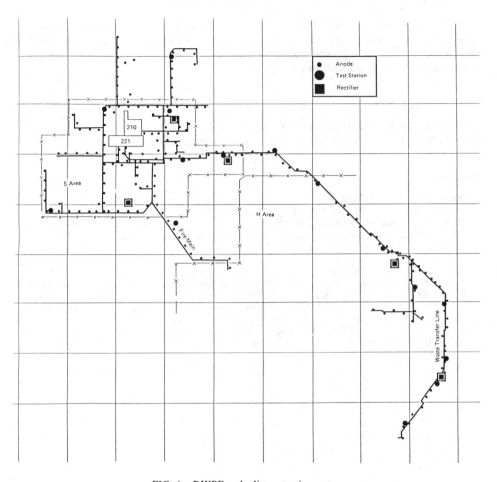

FIG. 6—*DWPF cathodic protection system.*

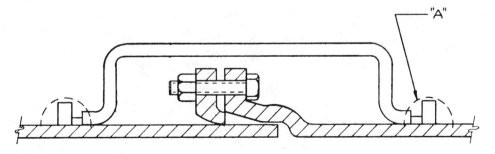

FIG. 7—*Cable jumper across mechanical joint on ductile iron pipe. Cable is thermit welded to the pipe at "A" and field coated with a coal tar mastic.*

The design may appear to be excessive. However, public safety and environmental concerns must be considered and piping containing high-level radioactive waste must be maintained at the highest integrity. In the present design, hopefully all of the factors have been recognized and precautions taken to avoid possible problems.

Acknowledgment

The information contained in this article was developed during the course of work under Contract No. DE-AC09-76SROOOOl with the U.S. Department of Energy.

References

[1] Karol, R. H., *Soils and Soil Engineering*, Chap. 7, Prentice-Hall, Englewood Cliffs, NJ, 1960.
[2] Miller, F. P., Foss, J. E., and Wold, D. C., "Soil Surveys: Their Synthesis, Confidence Limits, and Utilization for Corrosion Assessment of Soils," *Underground Corrosion, ASTM STP 741*, E. Escalante, Ed., American Society for Testing and Materials, Philadelphia, 1981, pp. 3–23.
[3] Peabody, A. W., *Control Pipeline Corrosion*, National Association of Corrosion Engineers, Houston, TX, 1967, p. 28.
[4] Kelley, W. P., *Cation Exchange in Soils*, Reinhold, New York, 1948, p. 125.
[5] Romanoff, M., *Underground Corrosion*, NBS579, National Bureau of Standards, U.S. Government Printing Office, Washington, DC, 1957.
[6] Colquhon, D. J. and Johnson, H. S., Jr., "Tertiary Sea-Level Fluctuation in South Carolina," *Palaeogeography, Palaeoclimatology, Palaeoecology*, Vol. 5, 1968, pp. 105–126.
[7] Corbett, R. A., "Cathodic Protection as an Equivalent Electrical Circuit," *IEEE Transactions on Industry Applications*, Vol. IA-21, No. 6, 1985, pp. 1533–1537.

James B. Bushman[1] and Thomas E. Mehalick[2]

Statistical Analysis of Soil Characteristics to Predict Mean Time to Corrosion Failure of Underground Metallic Structures

REFERENCE: Bushman, J. B. and Mehalick, T. E., "**Statistical Analysis of Soil Characteristics to Predict Mean Time to Corrosion Failure of Underground Metallic Structures,**" *Effects of Soil Characteristics on Corrosion, ASTM STP 1013,* V. Chakar and J. D. Palmer, Eds., American Society for Testing and Materials, Philadelphia, 1989, pp. 107–118.

ABSTRACT: In past years, the corrosion engineer has not been able to assess the influences of various soil characteristics on the rate of corrosion of buried ferrous metal structures. In 1979, the application of multivariate statistical analysis techniques to evaluate a broad range of measurements of soil and structure data was first used successfully to predict a mean time to corrosion failure (MTCF) for underground storage tank systems (USTSs). More recently, similar procedures have been developed to determine the MTCF for discrete sections of cast iron water pipelines (CIWPs). This paper will summarize the methodologies used to predict MTCF of both USTSs and CIWPs. It is anticipated that these methodologies may be more broadly applied to other buried metal structures, including oil and gas transmission and distribution systems, steel-reinforced concrete pipelines, etc.

KEY WORDS: corrosion, mean time to corrosion failure, underground storage tanks, cast iron water pipelines, multiple regression analysis, statistical analysis

Because corrosion is an electrochemical process, the corrosion rate experienced by any buried or submerged metallic structure is related to Ohm's law. The corrosion current is directly proportional to the voltage of the corrosion cell and inversely proportional to the resistance of the corrosion cell. This process can proceed in two distinct ways, with important consequences for the expected useful life of the structure.

The first of these two processes is uniform corrosion. This occurs when there are no anomalies on the surface or in the backfill of the buried structure and corrosion occurs uniformly over the structure's surface until the corrosion cells are polarized, resulting in a very low rate of corrosion. The second process, pitting corrosion, occurs when anomalies exist on the structure's surface or in the backfill in direct contact with the structure. Pitting corrosion is concentrated on a small portion of the surface area of the structure and is not uniform. Consequently, high rates of corrosion continue until perforation of the structure ultimately results.

For containment vessels such as pipes or tanks, the problem of pitting corrosion is of significant concern. In general, buried structures are more often subjected to pitting corrosion attack rather than uniform corrosion. Numerous corrosion investigations by both the authors and their associates indicate that the corrosion attack typically occurs on 5 to 10% of the structure's

---

[1]Vice president of Research and Development, Corrpro Companies, Inc., Medina, OH 44258.
[2]Manager, Special Projects, Corrpro Companies, Inc., Medina, OH 44258.

surface area [1–4]. This pitting attack significantly increases the rate of corrosion and is the problem to which the corrosion engineering community primarily directs its attention.

Measuring Corrosion-Inducing Influences

When evaluating the form, amount, and rate of corrosion on buried structures, corrosion engineers are compelled to rely on obtainable measurements short of actual excavation and examination of the underground structures.

For a number of years, corrosion engineers have used structure-to-electrolyte as well as other electrical potential measurement techniques to analyze corrosion patterns on underground pipelines [5]. These methods enabled the engineer to determine where active corrosion was occurring, but did not allow him to assess the rate and, therefore, time to failure. Thus, his only recourse was to apply corrosion mitigation techniques at each site where active corrosion was detected.

In the late 1950s and early 1960s, independent consultant Gordon Scott began to evaluate the use of soil resistivity to predict the relative corrosivity of the environment for steel-reinforced concrete pipelines [6, 7]. Studying the resistivity of soils, he determined that the resistivity data were normally distributed if the logarithm of resistivity is used in the analysis. Later researchers, including Husock and Steinert, extended his work to evaluation of the probability of corrosion failures versus the logarithm of soil resistivity to steel and gas transmission pipelines [8–10]. Their techniques enabled the engineer to compare approximately one-mile-long sections of transmission pipeline and to develop a priority schedule for application of cathodic protection for corrosion control on a programmed basis. Thus, those locations with the highest probability level of encountering low-resistivity values would be cathodically protected first, with action for other sections being deferred until later years. These techniques were expanded to incorporate the use of structure-to-electrolyte potential measurements which have provided some greater accuracy in their pipeline section selection process.

Subsequent investigations have disclosed additional environmental or structural variables which affect the rate of corrosion and are summarized in Fig. 1 [11–16].

It is important to note that one can observe considerable variance in the measurements of corrosion-inducing variables. A recent report by Kroon on data obtained at 2894 underground storage tank facilities (7590 tanks) ranging in age from 1 to 31 years demonstrated significant variance in the measurements obtained [17]. A summary of the variations in soil characteristics that he reported are provided in Tables 1 through 8.

Environmental and Structural Factors Impacting Rate of Corrosion

Some of the factors associated with the soil environment in which metallic structures are buried which can impact on the corrosion rates experienced include:

1. Moisture content.
2. Resistivity.
3. Permeability.
4. Chloride ion content.
5. Sulfide ion content.
6. Sulfate ion content.
7. Presence of corrosion-activating bacteria.
8. Oxygen content.
9. pH.
10. Total hardness and hardness as $CaCO_3$ of soil moisture.
11. Stray d-c currents.

Variable	Influence
Moisture Content	Under most circumstances, soil moisture content is directly related to corrosion rates.
pH	The lower the pH (below a neutral value of 7.0), the greater will be the corrosion rate. As pH increases above 10, conditions become increasingly more passive.
Sulfides	The presence of sulfides is often an indictor of sulfate reducing bacteria (SRB's). These bacteria can shift the pH in the acidic direction, causing accelerated corrosion.
Chlorides	The presence of increasing concentrations of chloride ions lowers the resistivity of soil and water and acts as a cathode depolarizer. Thus, increasing concentrations of chlorides in the soil moisture will increase the corrosion rate.
Hardness	The higher the water hardness, as expressed by the concentration of calcium carbonate, the lower the corrosion rate of steel and other ferrous metal structures.
Resistivity (Conductivity)	The lower the conductivity of the soil, the lower the corrosion rate.

FIG. 1—*Chemical impact on rate of corrosion.*

TABLE 1—*Soil resistivity (mean value at site).*

Ohm-Centimeters	% Occurrence
<3 000	16.1
3 000 to 9 990	37.1
10 000 to 19 900	20.7
20 000 to 49 900	16.5
>50 000	9.6

TABLE 2—*Soil conductivity (mean value at site).*

Micromohs	Ohm-cm	% Occurrence
<20	>50 000	0.1
20 to 49	50 000 to 20 000	2.6
50 to 99	20 000 to 10 000	10.7
100 to 999	3 000 to 1 000	21.2
>999	>1 000	6.9

TABLE 3—*Variation in soil resistivity (maximum-minimum per boring).*

Ohm-Centimeters	% Occurrence
<3 000	17.4
3 000 to 9 990	25.3
10 000 to 19 900	19.4
20 000 to 49 900	20.2
>50 000	17.7

TABLE 4—*Variation in tank-to-soil potential (maximum-minimum per boring).*

Millivolts	% Occurrence
<20	25.3
20 to 39	27.8
40 to 59	17.7
60 to 79	11.2
>80	18.0

TABLE 5—*Soil pH (mean value at site).*

pH	% Occurrence
<4.0	0.1
4.1 to 6.0	5
6.1 to 7.0	10
7.1 to 8.0	27
8.1 to 10.0	55
>10.0	3

TABLE 6—*Soil moisture content (mean value at site).*

% Dry Weight	% Occurrence
<5.0	15.2
5.0 to 9.9	48.3
10.0 to 14.9	27.0
15.0 to 19.9	7.6
20.0 to 24.9	1.2
>25	0.7

TABLE 7—*Chloride ion concentration (maximum at site).*

Milligrams per Liter	% Occurrence
0 to 9	43.4
10 to 19	19.9
20 to 49	16.3
50 to 99	7.8
>100	12.6

TABLE 8—*Sulfide ion concentration (maximum at site.)*

Milligrams per Liter	% Occurrence
=0.000	66.2
0.001 to 0.999	23.9
1.000 to 4.999	8.0
>5.000	1.9

Structure factors which can give impact on the corrosion rates experienced include:

1. Metallurgy.
2. Type and condition of coating.
3. Diameter.
4. Length.
5. Temperature.
6. Movement or vibration.
7. Stress level and stress distribution.
8. Presence of dissimilar metals.

Application of Mean Time to Corrosion Failure Analysis for Extensive Buried Structures

In the late 1970s, Warren Rogers identified that greater precision of measurement and more thorough analysis of corrosion-inducing variables could be conducted in order to determine the MTCF and the probability of leak for underground storage tanks. Using data gathered from the backfills of an extensive number of both failed and sound systems, he was able to develop a model to determine MTCF for USTSs using comprehensive measurements obtained at the site. This model has subsequently been applied at over 22 000 sites in North America and refined and verified based upon observations recorded following the excavation of tank systems previously evaluated [18].

Given that the feasibility of such an analysis had been established, the authors recently developed a sampling and research design to determine MTCF for pipelines in Western Ohio and Colorado. Each of these pipeline systems was experiencing an accelerated corrosion rate. However, the owners of these systems did not have funds budgeted for cathodically protecting these pipelines in their entirety over the short term. In fact, comprehensive cathodic protection of these particular systems would have been quite expensive.

In these particular cases, the costs of applying corrosion mitigation techniques were more significant than might be expected to be incurred for protecting transmission lines. In order for such a system to be protected in its entirety, it must be made electrically continuous. However, the systems under investigation consisted of bell and spigot joint, rubber-gasketed cast iron pipe; "bell hole" excavation of the pipe at each joint and thermite bonding of copper electrical continuity cables across each joint would be required to establish electrical continuity for comprehensive cathodic protection installation.

Therefore, a primary goal of the study was to apply failure analysis to determine MTCF for 100-ft increments along the entire length of each pipeline. Application of such a procedure would enable the pipeline operators to prioritize sections of piping for selective upgrading. For these two studies, the sampling plan developed was to bore holes at approximately 100-ft intervals along the pipeline as well as at each point where a known corrosion failure had occurred. In order for statistical analysis techniques to be useful, it was necessary to conduct measurements where corrosion failure had not occurred as well as points where pitting corrosion had been determined to be operative. Such a sampling plan was called for in order for valid data censoring of the corrosion-inducing variables.

The holes were drilled approximately 2 ft deeper and 6 in. off to the side of the pipeline. In-situ soil resistivity and electrical potential measurements were gathered at 1-ft intervals in each hole. In addition, soil samples were obtained just above, at the midpoint, and just below pipe depth using split spoon sampling techniques. Subsequent laboratory analysis of the samples was conducted for measurement of corrosion-inducing variables.

Once data were obtained according to the sampling plan, a site-specific corrosion model was developed for each of these two systems. Each data set was examined for fundamental mathematical characteristics such as the average, standard deviation, minimum, maximum, etc. All variables were then analyzed using the multivariate, linear, and nonlinear regression analysis techniques to develop a mathematical model for predicting the MTCF. The multiple regression analysis model developed is based on the general form:

$$Y = B_o + B_1X_1 + B_2X_2 + \ldots B_kX_k + e$$

where

Y = the dependent variable (for example, MTCF in years for each tested CIWP location),

X_k = each independent variable which impacts the MTCF (for example, soil resistivity, moisture content, etc.),

B_k = coefficient developed for each independent variable based on the relative contribution of each variable on the MTCF,

B_o = constant or Y intercept, and

e = random error possessing a normal probability distribution and having a mean equal to zero with a constant variance.

In the model, the response variable is the logarithm of known corrosion failure times. Independent variables are assumed to be the values of the corrosion-inducing variables obtained from the on-site measurements. However, given the limited sample size attendant with these two evaluations, some transformation of these parameters was required.

Based on this modeling, the distribution of failure times as related to the independent variables was established. Given that a normal distribution of failure times was achieved as a result of this analysis, the MTCF for this evaluation is established when the probability of a corrosion failure is 0.5.

As a result of such analysis, a model was developed which provided predicted MTCF for each designated interval of the two pipeline systems under investigation. MTCF can be interpreted as the average age at which each section of pipeline would experience perforation due to pitting corrosion.

Data plots were then generated which matched predicted MTCF against actual pipe age. The results of this analysis are provided in Fig. 2. In addition, this figure shows the actual pipe ages when corrosion failure occurred and the MTCF at each point based upon the values of the corrosion-inducing variables at the site. Those sites where the predicted failure age approaches or dips below actual pipe segment age are the locations designated for immediate management attention (Fig. 3). Subsequent activities could be anticipated to be prioritized based upon the amount to which predicted MTCF exceeds actual pipe age.

However, it should be recognized that these values shown are specific to the sites evaluated and are particular to this limited sample.

In order to better appreciate the inability of using any single soil characteristic to predict the MTCF for these cast iron water lines, data scatter plots (Figs. 4–8) were generated for several of

Location (in feet)	Actual Leak (Years After Installation)	Predicted MTCF
6140	18	16.8
6153	18	16.5
6311	6	10.8
6444	16	16.8
8769	30	29.3
9063	29	28.5
9100	29	29.9
9282	32	28.2
9319	32	31.8
9900	28	28.6
9917	28	31.6

FIG. 2—*Predicted MTCF versus actual leakage.*

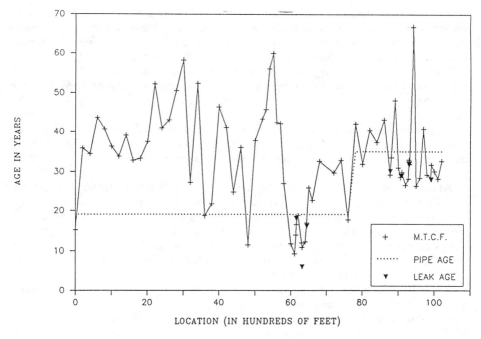

FIG. 3—*MTCF versus known leaks: cast iron water line.*

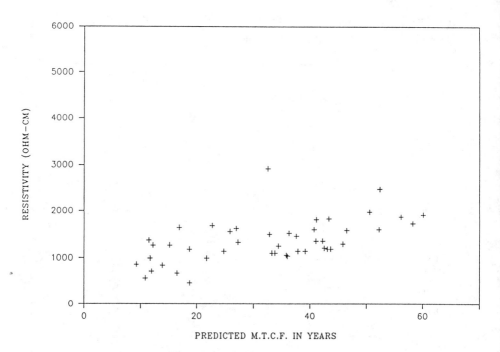

FIG. 4—*MTCF versus in-situ soil resistivity: cast iron water line.*

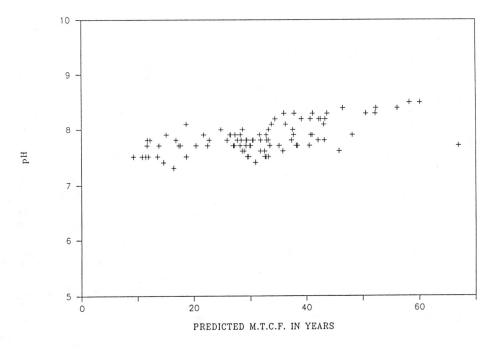

FIG. 5—*MTCF versus pH: cast iron water line.*

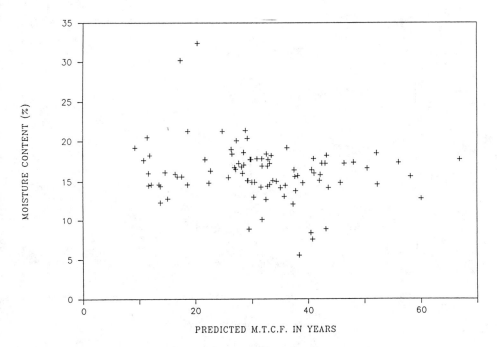

FIG. 6—*MTCF versus moisture content: cast iron water line.*

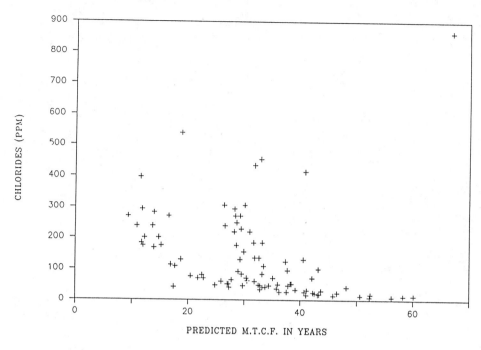

FIG. 7—*MTCF versus chlorides: cast iron water line.*

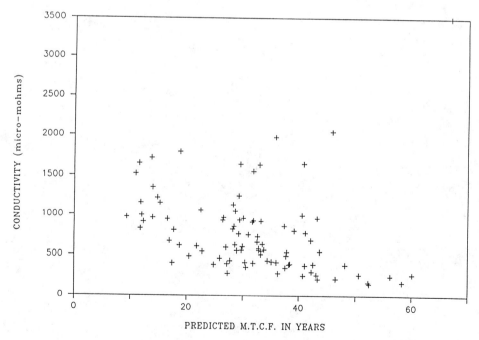

FIG. 8—*MTCF versus soil sample conductivity: cast iron water line.*

the variables used in the model versus the predicted MTCF at each test location. No individual graph plotted in this manner would provide the corrosion engineer with insight into how each of these variables relates to the MTCF.

Conclusions

1. Statistical analysis techniques provide an extremely powerful management tool for economic evaluation of the application of corrosion mitigation and replacement programs for USTSs and CIWPs.

2. Models for determining mean time to corrosion failure of long pipeline-type containment structures are in an initial stage of development and, hence, specific to the project. As more data become available, more comprehensive models may be developed which can be generalized to geographically dispersed, but similar structures.

3. More extensive measurements will be required to develop an adequate protocol for determining optimum sampling plans for measuring corrosion-inducing variables in the backfills of long pipeline-type containment structures so that more comprehensive analysis can be conducted.

4. The development of reliable procedures for determining the MTCF for discrete portions of ferrous pipeline structures will enable the corrosion engineer to plan and budget for both phased replacement and upgrading programs through the application of cathodic protection for corrosion control.

References

[1] Romanoff, M., "Underground Corrosion," N.B.S. Circular 579, U.S. Government Printing Office, Washington, DC, 1957.

[2] "External Corrosion—Introduction to Chemistry and Control," *AWWA Manual No. M27*, American Water Works Assn., Denver, CO, 1987.

[3] Meyers, J. R., "Fundamentals and Forms of Corrosion," Civil Engineering School of the Air Force Institute of Technology, Wright-Pattern Air Force Base, OH, October 1974.

[4] Uhlig, H. H. and Revie, R. W., *Corrosion and Corrosion Control*, John Wiley & Sons, New York, 1985, pp. 178–186.

[5] Husock, B., "Use of Pipe-To-Soil Potential in Analyzing Underground Corrosion Problems," *Corrosion*, August 1961.

[6] Scott, G. N., "The Distribution of Soil Conductivities and Some Consequences," *Corrosion*, August, 1958.

[7] Scott, G. N., "Distribution of Soil Conductivity and its Relation to Underground Corrosion," *Journal AWWA*, No. 52, March 1960.

[8] Husock, B., "A Method for Evaluating Corrosion Activity on Bare Pipelines," presented at the National Association of Corrosion Engineers Annual Convention, Anaheim, CA, March 1973, NACE, Houston, TX.

[9] Husock, B., "A Statistical Probability Method for Soil Resistivity Determination," IEEE Paper No. A79-077-9, IEEE, New York, 1979.

[10] Steinert, W. W., "Use of Extreme Value Statistical Theory in the Evaluation of Corrosion Problems," presented at the Northeast Regional Conference, National Association of Corrosion Engineers, October 1975 in Hartford, CN.

[11] Schaschl, E., and Marsh, G. A., "The Effect of Dissolved Oxygen on Corrosion of Steel and on Current Required for Cathodic Protection," *Corrosion*, April 1957.

[12] Logan, K. H., Ewing, S. P., and Denison, I. A., "Soil Corrosion Testing," *Symposium on Corrosion Testing Procedures*, ASTM, 1937.

[13] Logan, K. H., "Engineering Significance of National Bureau of Standards Soil-Corrosion Data," *Journal of Research*, NBS, Washington, DC, 1939.

[14] "Procedures for Evaluating Pipeline Replacement, a manual," East Ohio Gas Company, Cleveland, OH, 1979.

[15] Kumar, A., Bergerhouse, M. and Blyth, M., "Implementation of a Pipe Corrosion Management System," Paper No. 312, National Association of Corrosion Engineers Annual Conference, San Francisco, CA, March 1987, NACE, Houston, TX.

[16] Munn, R. S., "Microcomputer Corrosion Analysis for Structures in Inhomogeneous Electrolytes," Paper No. 50, National Association of Corrosion Engineers Annual Conference, Houston, TX, March 1986.
[17] Kroon, D. H., "Integrity Assurance Program for Underground Storage Tank Systems," presented at the National Association of Corrosion Engineers National Conference, San Francisco, California, March 1987 and reprinted as Paper No. CP-13, Corrpro Companies, Inc., Medina, OH, April 1987.
[18] Rogers, W., report by Special Task Force on Underground Storage Tanks, Appendix E, Petroleum Council for the Protection of Canadian Environment, Toronto, 1980.

Bibliography

Brown, M. B., ed., *Biomedical Computer Programs,* University of California Press, Berkeley, 1977.
Draper, N. and Smith, H., *Applied Regression Analysis,* Wiley, New York, 1966.
Kleinbaum, D. and Kupper, L., *Applied Regression Analysis and Other Multivariable Methods,* Duxbury, North Scituate, MA, 1978.
Ray, A. A., ed., *SAS User's Guide: Statistics,* SAS Institute, Inc., Cary, NC, 1982.
Younger, M. S., *A Handbook for Linear Regression,* Duxbury, North Scituate, MA, 1979.
SPSS Users Guide, SPSS/PC and Software Products, Chicago, IL, 1987.

K. P. Fischer¹ and O. R. Bryhn²

Corrosion and Corrosion Evaluation of Superficial Sediments on the Norwegian Continental Shelf³

REFERENCE: Fischer, K. P. and Bryhn, O. R., "**Corrosion and Corrosion Evaluation of Superficial Sediments on the Norwegian Continental Shelf,**" *Effects of Soil Characteristics on Corrosion, ASTM STP 1013,* American Society for Testing and Materials, Philadelphia, 1989, pp. 119–132.

ABSTRACT: In-situ corrosion tests and laboratory corrosion studies have been performed on superficial marine sediments from water depths down to 500 m. The corrosivity of the superficial sediments on the Norwegian continental shelf is generally low, being in the range 1 to 50 μm/year. For most sediments which have a glacial or glaciomarine origin and have a low organic content, the corrosion rate will be in the range of 1 to 20 μm/year. However, in some limited areas where the sediments contain a high amount of partly decomposed organic material, the corrosion rate can be as high as 360 μm/year. It is considered that the sulfide environment created by the reduction of the biogenous material will decide the corrosivity of these marine sediments. Based on the studies performed on the various marine sediments found on the shelf, a reasonable evaluation of the corrosivity can now be based on the soil description and the geotechnical properties. However, more detailed electrochemical studies are necessary for an accurate determination of the corrosivity rate. A methodology for the evaluation of the corrosivity of steel in marine sediments is presented.

KEY WORDS: corrosion, marine sediments, steel, polarization, sulfides, deep waters

The placing of permanent offshore structures on the Norwegian continental shelf has increased the need for a more exact knowledge of all aspects of corrosion and corrosion control. The major difficulty in a systematic evaluation of the corrosivity of undisturbed anaerobic sediments is the fact that the actual corrosion mechanisms are unclear. Due to the complexity of designing and performing in-situ corrosion control tests in deep water areas, few published data are available. Investigations performed in the Pacific Ocean off the coast of Port Hueneme, in water depths of about 760 and 1830 m, gave corrosion rates for steel of about 20 to 40 μm/year for long-term exposure of 1 to 3 years [1]. Values up to 75 μm/year have been reported from exposures at shallower depths (20 to 100 m) in the Gulf of Mexico [2].

The marine sediments are mainly anaerobic and the activity of sulfate-reducing bacteria (SRB) has been considered to be the main cause for free corrosion [3,4]. As will be discussed subsequently, the sulfide environment can initiate or increase the corrosion rate. Thus, the corrosivity could ideally be predicted based on the complete data on the environment and the nutritional situation for the SRB. King has made such a scheme for corrosion prediction based on sediment type, organic content, water depth, seawater content of nitrogen and phosphorous,

¹MARINTEK A/S, P.O. Box 173, 3201 Sandefjord, Norway.
²Norwegian Geotechnical Institute (NGI), P.O. Box 40, Tasen, 0801 Oslo 8, Norway.
³This paper has been presented at the 10th Scandinavian Corrosion Congress, Stockholm, Sweden, 2–4 June 1986.

and temperature [4]. In the application of this scheme for the North Sea area, King predicted a low corrosivity for the main part of the Norwegian sector of the North Sea basin. However, along the coast a risk of SRB-induced corrosion can be expected partly due to the presence of mud (Norskerenna) and typically high phosphorous content along the south coast of Norway.

In the design of CP (cathodic protection) for buried structures on the Norwegian continental shelf, the assessment of the corrosivity has been based on experience from other areas of the world. It is the purpose of the present paper to present the data obtained from sediments on the Norwegian Continental Shelf and to discuss methods applicable for the evaluation of the corrosivity of seabed sediments.

Theoretical Aspects of Corrosion in Sediments

Marine sediments represent a complex and heterogeneous environment containing both inorganic and organic constituents with seawater as the original pore water. However, the reactions occurring in the sediment may result in a change in the equilibrium and the pore water composition with time. Often a decline in the sulfate content of the pore water is observed due to the activity of the SRB. Oxygen will in most cases only diffuse just a few centimeters into sea bottom, and thus the deeper sediments will be anaerobic with a redox potential of -100 mV (SHE) or more negative [3]. Corrosion in these environments, when considering undisturbed sediments, will exclude oxygen as a cathodic reagent. For this reason microbial corrosion and the effects of sulfides have been regarded as major causes of corrosion in the marine sediments. In the sediment a slow decomposition of plant and animal residues will result in the formation of carbon dioxide, ammonia, and phosphate. The decomposition of proteins to amino acids and further to sulfides or sulfates is part of the sulfur cycle in the sediments [3]. The bacterial reduction of sulfate to sulfide is a critical part of the sulfur cycle in the marine sediments. The formation of sulfides leads to the precipitation of ferrous sulfide (giving a black or grey color to the sediment) and also produces free H_2S in the sediment. In general it is expected that the importance of SRB reduction will decrease with increasing water depth and distance from land, due to the lower amount of organic sedimentation [5].

Certain aspects of the mechanism related to SRB-type corrosion of ferrous metals are still unclear [6, 7]. These uncertainties are related to the influence of the sulfides formed during the SRB metabolism as well as the properties of the organic and inorganic constituents of the environment. Thus, as listed below, a variety of factors can influence the corrosion in a given sediment or soil system [6, 7]:

1. Corrosion will be caused by sulfate-reducing bacteria. The rate-determining factor can be the microbial ability to utilize hydrogen and/or the influence of the corrosive substances produced by the bacterial metabolism.

2 The influence of precipitated ferrous sulfides on the cathodic reaction.

3. The influence of the sulfides on the anodic reaction.

4. Formation of protective or nonprotective types of ferrous sulfides on the cathode; for example, a high ferrous content in the environment will increase the possibility of the formation of a nonprotective film.

5. Formation of galvanic cells due to the presence of ferrous sulfides.

The protective properties of the ferrous sulfide films change with time [8]. Often a thin protective film will form initially, but after some days or even months this film can crack, and subsequently the rate of corrosion will increase significantly. In such a case the corrosion rate in most cases remains at a high value. However, as given above, the influence and the properties of these ferrous sulfide films are unclear and difficult to predict. Barker Jorgensen found that the sulfate reduction rates in marine sediments vary from less than 1 nmol SO_4^{2-} cm^{-3} day^{-1} in deep sea areas to about 100 nmol SO_4^{2-} cm^{-3} day^{-1} in coastal regions [5]. If all of the sulfide produced resulted in corrosion, the equivalent corrosion rates would range from less than 1 to about 5 μm/year. Similar investigations of SRB metabolism based on sulfide production have been published by Ramm and Bella [9]. From their studies on core samples and field tests, the sulfide production would be equivalent to corrosion rates ranging from less than 1 and up to about 80 μm/year. For studies with pure SRB culture growths, the sulfide production rates would be equivalent to corrosion rates as high as about 400 μm/year [9]. With respect to the importance of depth below sea level, a study by Willingham and Quinby [10] showed that the corrosion of steel in an SRB culture at a hydrostatic pressure of 200 bar was greater than a similar culture test at 1 bar.

The importance of sulfide chemistry in the corrosion processes can be evaluated in the context of polarization properties of steel in such an electrolyte. As an approximation, the polarization properties of steel in a buffered aqueous sulfide solution can be utilized. In an investigation by Bolmer [11], a cathodic Tafel slope of about 70 mV/decade was found in a basic solution of 10^{-2} M HS^-, 0.1 to 0.3 atm H_2S, which increased to about 120 mV/decade at 10^{-3} M HS^-, 0.06 to 0.3 atm H_2S. In a more recent study by Morris et al. [8] at pH 3, it was found that the anodic Tafel gradient was about 40 mV/decade. They found a cathodic Tafel gradient of about 120 mV/decade and that the occurrence of a hydrogen limiting diffusion current disappeared in the presence of sulfide. It was reported by Morris et al. that the H_2S environment did not change the Tafel slopes of the anodic and cathodic processes. Further, they found that the potentials in such sulfide systems are shifted to more negative values mainly due to a decrease in the reversible potential of the steel.

Soil Conditions on the Shelf Area

The mapping and investigations of the shelf area has been intense during the last ten years (Fig. 1). Detailed maps of the shelf areas have been published by IKU (Continental Shelf Institute), and good review articles of the geology of the superficial sediments have been presented by Loken [12] and Gunnleiksrud [13].

In the northern North Sea (57 to 62°N), the sea floor generally consists of sand and gravel at the plateau, and clay and silty clay in the Norwegian Trench. In the Norwegian Trench the clay has a high water content and low resistivity (50 to 100 ohm-cm). The SRB counts vary from 0 to 105 per gram wet sediment. The sulfide content is generally low in the top 2 m, but increases with depth. The pH is generally 7.5 to 8.3. Both the eastern and western slopes consist of more sandy silty clays covered with a thin sand deposit. Here the resistivity of the clay is in the order of 70 to 120 ohm-cm with a pH between 7.0 to 8.0. Nearly black sediments sometimes with a strong smell of H_2S have been found, but sediments almost free from sulfides may also occur. The occurrence of H_2S is especially marked in the sand deposit at the plateau where black iron sulfide only occurs at great depths. For the whole area the organic carbon content seldom exceeds 0.5%. In the region 62 to 67°N, only scattered relevant data are available. Generally the plateau consists of clay with dropstones or sand with gravel. Clay deposits become more and more dominant with increasing depth. The few results which exist indicate a certain occurrence of sediments containing some sulfides. North of 67°N the same variation in the deposit as described above occurs. Most of the samples for corrosion evaluation are sandy silty clay, some of them with black spots of iron sulfides. The pH usually varies between 7.0 and 8.3. West of Tromsoflaket, the location at 500 m depth consisted of layered sand, clay, and organic debris interrupted by old iceberg plow marks infilled with soft clay. The occurrence of SRB varies from 0 to 106 counts per gram. The total sulfide content may vary considerably; from earlier studies the sulfide content was found to be in the range 0 to 1% (as FeS) [14]. On occasions the smell is more pungent, indicating the present of amines. The amount of organic carbon varies between 0 and more than 2% (as C).

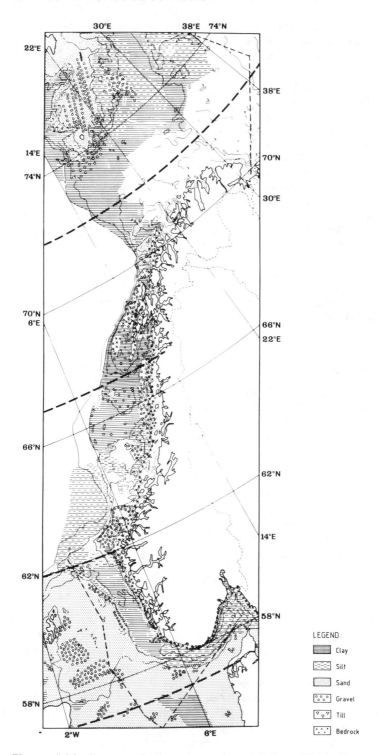

FIG. 1—*The superficial sediments on the Norwegian continental shelf (compiled by Norwegian Geotechnical Institute from data from Norwegian Polar Research Institute, Continental Shelf Institute, and University of Bergen Department for Geology).*

Methodology

The majority of tests in-situ were concerned with free corrosion exposures based on weight loss. Here steel samples (Norwegian Standard NS12153 Grade 52-3, which is comparable to BS4360 Grade 50C) grit blasted to Sa 3 (Swedish Standard SIS 055900) were used (coupon size 100 by 150 mm). The exposure period was about six months. During the field survey, sediment samples were collected from the seabed down to a depth of 5 m into the sediment (superficial sediments). The samples were mainly core samples (undisturbed soil), but grab samples (disturbed soil) were also taken. These samples were utilized for studies in the laboratory. In the soil evaluation, the following parameters were determined: resistivity, SRB counts, pH, and geotechnical and geochemical studies. The samples were stored at 7°C and at nearly 100% relative humidity, except during the transportation period. All long-term laboratory tests were carried out at 7°C. However, the resistivity was measured at 20°C. From earlier studies it has been found that the salinity of the pore water of superficial sediments will be close to the salinity of the seawater [14]. Thus, the resistivity of the sediments can be calculated to the in-situ temperature of 7°C as a temperature effect. At a salinity of 3.43%, the resistivity (ohm-cm) at 20°C will be 27% lower than at 7°C.

Galvanostatic polarization tests with a three electrode probe designed by NGI have been described earlier [15]. In some cases a three electrode polarization technique was applied with the use of the three electrodes individually mounted in the soil sample. The reference electrode was placed at about 0.5 to 1 cm from the working electrode. No corrections were made for the IR drop but the low currents used should limit the IR drop to less than 10 mV. The polarization curves were obtained soon after the exposure of the probe, and repeated experiments were carried out at various times of up to 20 days exposure. At a polarization in excess of about 30 mV, a Tafel line was generally found, but in some cases a Tafel gradient did not appear for the cathodic polarization (diffusion overvoltage). The corrosion rates reported were determined from the intercept of this cathodic Tafel line with the corrosion potential (Tafel line extrapolation). All potentials refer to the Ag/AgCl (seawater) electrode.

As a result of the past studies related to corrosion and corrosivity evaluation of the sediments on the continental shelf, the following scheme of evaluation from A to D is considered applicable:

Point A: Soil Description:

1. Evaluate geological survey reports.
2. Sample description (for example, clay, silt, or sand, color, smell, homogeneous or layered, sample collection date and storage conditions prior to examination).

Point B: Soil Corrosion Parameters:

1. Resistivity.
2. Test for FeS (for example, with dilute HCl).
3. Geochemical data (for example, organic content, sulfide content, pore water chemistry).

Point C: Electrochemical Parameters:

1. Free corrosion potential.
2. Polarization studies and time.
3. Free corrosion weight loss, long-term exposure.

Point D: Special Investigations:

1. SRB and microbiological analysis.
2. Sulfide production and precipitation.
3. Synergetic effects: clay-SRB-sulfide Interactions.

The following comments can be made to this evaluation sequence: The soil description (A) is a most valuable basis for a corrosivity evaluation. This soil investigation should be done as fast as possible after a sample recovery to ensure that no drying or oxidation occurs. If the sample has a dark color (grey or black) the sample should be tested for the presence of FeS by adding a dilute HCl (2 to 5% solution) to the clay. The smell of H_2S will indicate the presence of sulfides. It is here important to remember that a negative result in this test will not exclude the possibility that sulfides may be present or may have been present. A reasonable prediction of the corrosivity can be based on points A and B if the soil description and geotechnical/geochemical data are similar to sediments which have been tested earlier. However, for a more exact evaluation of the corrosivity, electrochemical investigations as detailed under point C should be performed. As discussed earlier, mainly anaerobic and sulfate-reducing bacteria will be active in the marine sediments. However, the traditional culture growth of SRB (counts of SRB per gram wet sediment) have shown no correlation with the corrosion rates. Thus, while the SRB count can vary from 0 to 107 counts/gram in such a test, this result was not found to be directly related to the corrosion rates. Tests of SRB and other microorganisms should be considered as special investigations. These should be performed when the organic content of the sediment is especially high or if there are factors which indicate that their metabolism or characterization is critical for the corrosion process. The SRB counting can also be useful in evaluating the stage in the geochemical development of the sediment. The properties of the sulfides in the sediment at the steel/sediment interface are considered to be critical to the corrosion state.

Results

The results from the various in-situ and laboratory tests are summarized in Table 1. In Fig. 2 is given a plot of all the determined corrosion rate and resistivity data. The scatter of the data points is considerable. However, the tendency is that the high corrosion rates will mainly occur at the lowest resistivities, but there is no simple relationship between the resistivity and the corrosion rate. As can be seen, the corrosion rates have been found to vary from about 1 to 360 μm/year for soils with resistivity in the range 70 to 160 ohm-cm. The free corrosion potentials as determined in the laboratory studies varied between about -650 to -730 mV. As given in Fig. 3, the corrosion rates found by the galvanostatic method (cathodic Tafel line) were here from about 150 μm and down to less than 1 μm/year for the free corrosion potentials in the range -650 to -730 mV.

Corrosion Rates: In-Situ and Laboratory Tests

At a depth of 300 m in soft clay with low sulfide content, the in-situ corrosion rate after one year exposure was 22 to 47 μm/year (average 33 μm/year) in the northern North Sea. Further north (62 to 67°N) at a depth of 270 m at Haltenbanken, the in-situ corrosion rate was 30 μm/year in sandy clay after 75 days exposure. North of 67°N an exposure was carried out at a depth of 220 m at Tromsoflaket in a sandy, silty clay with black spots giving a corrosion rate of 19 μm/year over a period of 183 days. At 500 m depth west of Tromsoflaket (71 to 72°), three series of tests lasting about six months gave corrosion rates in the range 80 to 360 μm/year (average 128, 250, and 299 μm/year). These extremely high corrosion rates are considered to be due to the presence of very corrosive organic material in the sediment.

Free corrosion tests were also carried out on undisturbed soil samples in the laboratory. From

Table 1.   Properties and corrosion data for samples and

sediments on the Norwegian Continental Shelf

Location (Northern latitude sector)	Water depth m	Sample depth m	Sediment description (colour)	Resistivity ohmcm (at 20°C)	pH	E_{corr} mV vs. Ag/AgCl	Corrosion rate µm/year		Tafel slopes	
							Weightloss (months)	From cathodic Tafel line (h)	Anodic mV/dec	Cathodic mV/dec
Tromsøflaket (71-72°N)	220	0.7	clay,silty (olive grey)*	118	7.7	-	19** (6)	-	-	-
West of Tromsøflaket (71-72°N)	505	0.1	clay (brownish)	140	-	-768	-	10 (0.3)	32	120
West of Tromsøflaket (71-72°N)	503-505	0.05-0.15	clay,sand silty clay (grey-brown)	90-250	-	-	80-360** (6)	-	-	-
Nordvestbanken (67-71°N)	252	0.8	clay,sandy,silty (dark grey)	84	7.4	-726	50 (7)	150 (96)	48	120
Nordvestbanken (67-71°N)	292	1.35	clay (dark grey)	48	7.5	-672	-	110** (192)	24	120
Fugløybanken (67-71°N)	271	1.4	clay,sandy,silty (dark grey)	93	7.9	-683	45 (7)	42 (24)	30	118
Andfjorden (67-71°N)	321	0.3	clay,silty (dark grey)	50	7.7	-645	-	72 (120)	27	80
Haltenbanken (64-67°N)	270	0.1	clay,sandy	-	-	-	31** (2.5)	-	-	-
Møre coast (62-64°N)	399	2.0	clay,silty (grey)*	49	7.4	-690	-	133 (120)	24	118
Northern North Sea (62-64°N)	374	0.35	clay,silty olive grey	33	7.5	-645	-	128 (120)	53	120
Norwegian trench (57-62°N)	320	0.75	clay,silty (olive grey)	59	7.9	-690	-	9 (24)	45	140
Northern North Sea (57-62°N)	305	0.5	clay (grey)*	55	-	-	22-47**(12) 17(12)	-	-	-
Northern North Sea (57-62°)	300	4.1	clay (grey)*	76	7.5	-715	0.7 (8)	1.6 (80)	38	118
Northern North Sea (57-62°)	218	0.4	clay,silty (greyish brown)	57	7.1	-698	13 (8)	3.6 (54)	32	126
Northern North Sea (57-62°)	218	1.15	clay,silty,sandy (greyish brown)	70	8.2	-710	-	3	37	90
Northern North Sea (57-62°)	224	1.4	clay,silty,sandy (dark grey)	65	7.8	-716	1 (8)	2 (124)	28	120
Northern North Sea (57-62°)	310	1.15	clay,silty,sandy (dark grey)	80	7.9	-669	4 (8)	63*** (122)	34	65
Northern North Sea (57-62°)	143	1.55	clay,silty,sandy (dark grey)	86	7.5	-705	0.8 (8)	2.3 (127)	32	120
Northern North Sea (57-62°)	109	0.1	sand,silty*	-	-	-662	-	60*** (5600)	30	120
Northern North Sea (57-62°N)	109	0.1	sand,silty			-695		3.1 (4070)	29	120

Legend:

```
*:    Black spots
**:   In situ
***:  Increases with time
(h):  exposure period prior to the polarization
```

the northern part of the North Sea (57 to 62°N), a number of ground steel plates have been exposed to different types of soils from sandy silt to clay. The tests cover regions with water depths from 145 to 325 m. Here the corrosion rates are very low, only 0.5 to 4 µm/year with a tendency towards higher corrosion rates in soils containing iron sulfides and H_2S. From 67°N and northwards only a few experiments have been carried out in undisturbed dark grey sandy, silty clay taken from a depth of 250 to 270 m. Grit-blasted steel plates have been exposed in this

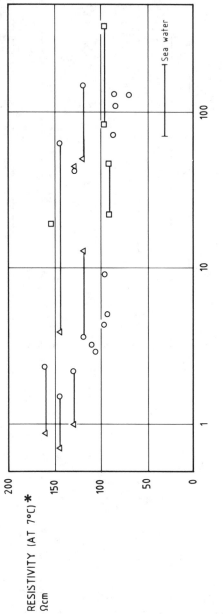

FIG. 2—*Soil resistivity and corrosion rates.*

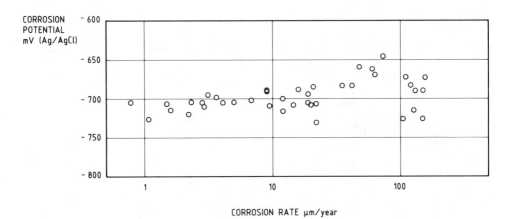

FIG. 3—*Free corrosion potentials and corrosion rates. The corrosion rates are based upon cathodic Tafel line extrapolations.*

clay for seven months, giving corrosion rates of 45 and 50 μm/year. Based upon this information there is reason to believe that ground steel plates represent the lowest corrosion rate while grit-blasted steel plates represent the maximum corrosion rate.

Corrosion Rates: Determined by Galvanostatic Polarization in the Laboratory

The corrosion rates will change with time due to the formation and properties of corrosion products and other deposits formed on the steel surface. To establish how the corrosion rates change with time, the galvanostatic polarizations were performed at different times after the exposure of the steel probe. As shown in Fig. 4, the corrosion rate would mainly decrease with time, but in some cases an increase was found. As shown in Fig. 4, the corrosion rate would reach relatively stable values after about 100 h. In Fig. 5 is shown a suite of galvanostatic polarization curves for a northern North Sea clay; here the corrosion rate would decrease to about 1 μm/year after a 500-h exposure period. While in Fig. 6 the similar suite of galvanostatic polarization curves for a different clay sample showed an increasing corrosion rate with time to reach a value of more than 50 μm/year after a 122-h exposure period.

From the southern part of the North Sea south of 57°N only one sample was tested. Here a deposit of silty sandy clay with black spots showed a corrosion rate of 4 μm/year. For the northern North Sea area the corrosion rates were between 1 and 3 μm/year for soil material with low sulfide content. While for those with a high sulfide content the corrosion rate was found to be about 50 to 70 μm/year. The cathodic Tafel slope was found to be between 65 and 140 mV/decade (Table 1). However, sulfide-rich soil samples showed a Tafel slope of about 65 mV. Further north, on the More Coast (62 to 64°N), samples of a dark grey silty clay with black spots from about 400 m depth were investigated. Here the polarization studies gave corrosion rates of about 130 μm/year. North of 67° four samples of dark grey clay from 250 to 350 m depth were tested. Here, the cathodic Tafel slope varied from 78 to 120 mV/decade. The corrosion rate obtained varied from 42 to 150 μm/year.

In general the corrosion rate, as estimated from the galvanostatic polarization method, reached a relatively stable value after about 100 to 1000 h exposure of the probe (Fig. 4). This corrosion rate value was found to be 2 to 3 times higher than the corrosion rate as determined by weight loss over a six-month exposure period.

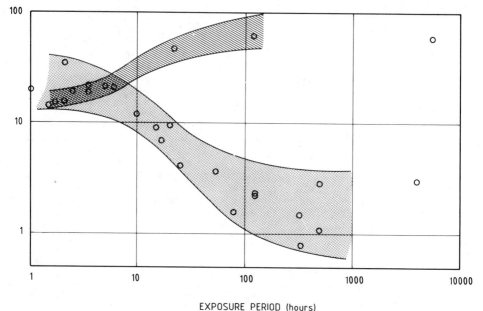

FIG. 4—*Corrosion rates based on cathodic Tafel extrapolation as a function of the probe exposure period.*

However, in the cases with a high activity of sulfides and/or when H_2S is present, the laboratory methods will often greatly underrate the corrosion rate as compared to an in-situ test. This is considered to be caused mainly by the oxidation of the sulfides or H_2S during handling or transport of the soil sample. Such an occurrence was considered to be the case for the in-situ test at 500 m depth (71 to 72°N) where the corrosion rate (weight loss) was 80 to 360 μm/year while the galvanostatic polarizations on the recovered soil sample only gave a corrosion rate of about 10 μm/year.

Cathodic Protection in the Seabed

On one occasion an in-situ cathodic polarization test was performed at 500 m west of Tromsoflaket. The anode-cathode area was 1:13 with a design current density of 180 mA/m² in a manner as described in an earlier study [16]. Unfortunately the instrumentation was damaged after 100 days exposure and the logger unit was lost. However, the recovered test unit revealed a cathode which was partly covered with a very dense calcareous deposit intermingled with mineral grains. The calculated average cathode current density based on weight loss of the Zn anode was 25 mA/m². As the design current density was very high, it can be expected that the CP maintenance current density would be at about 10 to 15 mA/m². Although this test was only partly successful, it has confirmed that a cathodic protection leading to a very protective calcareous deposit will be established in these clays at 500 m depth.

A similar cathodic polarization test was performed in the laboratory with a design current density of 160 mA/m². The soil sample here was a soft clay from the northern North Sea at a water depth of 300 m. This was a grey clay with black spots of ferrous sulfide. The SRB content

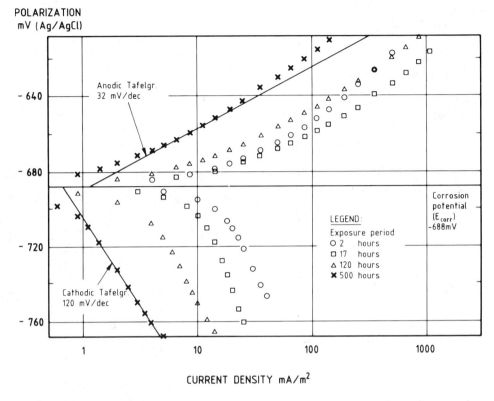

POLARIZATION
mV (Ag/AgCl)

Anodic Tafelgr.
32 mV/dec

- 640

Cathodic Tafelgr.
120 mV/dec

- 680

- 720

- 760

1   10   100   1000

CURRENT DENSITY mA/m²

LEGEND:
Exposure period
○ 2   hours
□ 17   hours
△ 120 hours
✗ 500 hours

Corrosion
potential
(E$_{corr}$)
-688mV

FIG. 5—*Northern North Sea clay: Polarization curves as a function of time. Normal case where corrosion rate decreases with time.*

was determined to be from 2×10^4 to 2×10^5 counts per gram. Here the steel was polarized instantaneously to -900 mV, and, subsequently, the potential decreased to -980 mV after one year. The final current demand was about 10 mA/m² and the average value was 17 mA/m². Thus from these present tests the cathodic protection current requirements are as expected in a marine sediment [*14*].

Discussion

Corrosion and Corrosion Rates on the Norwegian Continental Shelf

As given in Fig. 1, the superficial sediments (0 to 5 m depth into the sea bottom) can vary from gravel to a soft clay [*12,13*]. The geological origin of these sediments can vary significantly, but the majority of the Norwegian Continental Shelf is covered by glacial or glaciomarine sediments. They were deposited during the last glaciation (Weichslian Glaciation) lasting from 70,000 until 10,000 years before the present. In a limited number of areas, sediments formed under more temperate geological climate can be found. These recently formed sediments (younger than 10,000 years) are termed Holocene sediments and are often deposited in depressions or channels. Usually these sediments contain a much larger amount of biodegradable material, and the activity of the microorganisms (SRB) can be expected to be much higher than in the

POLARIZATION
mV (Ag/AgCl)

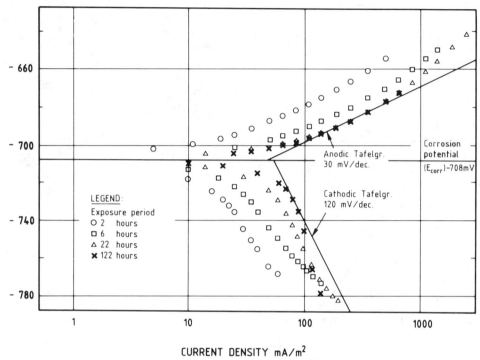

FIG. 6—*Northern North Sea clay: Polarization curves as a function of time. Here the corrosion rate increases with time, which is the exceptional case.*

sediments formed under the colder climate. These differences in the sediment properties will have to be considered in addition to the general trend of decreasing activity of SRB with increasing water depth and distance from land [5].

The results from the in-situ corrosion exposures performed have generally given very low corrosion rates. Thus, the corrosion rates (weight loss) to be expected in the major part of the Norwegian Continental Shelf will be in the range 1 to 50 μm/year. From what has been presented above, corrosion rates in the range 20 to 50 μm/year can be expected in the Holocene sediments. In the sediments which have a glaciomarine origin and contain usually a low amount of biodegradable material, the corrosion rates will be in the range 1 to 20 μm/year. The central North Sea plateau with its superficial sediments of glacial or glaciomarine origin will be an area with expected corrosion rates of 1 to 20 μm/year. However, in some depressions or limited areas on the plateau, the corrosivity will be higher and in the range 20 to 50 μm/year.

Further north of 62°N the corrosion rates will generally be in the low range 1 to 20 μm/year. Holocene material is very scarce and will only constitute a thin layer in most depression areas on the shelf. In this area of the shelf (100 to 500 m depth) there are extensive plow marks (up to 25 m deep, 250 m wide, and several km long) due to scouring by drifting icebergs at the end of the Weichslian Glaciation. Generally, the amount of Holocene material is also low in the deep parts of these channels [13].

It can be concluded that the corrosion rate of steel on the Norwegian Continental Shelf will in general be very low in most areas and 20 to 50 um/year in limited areas where Holocene sediments occur. However, in an area at 500 m depth (71° to 72°N), a sandy clay was found with a corrosion rate in the range 80 to 360 μm/year. In this case the sediment sample was very rich in organic material in an early stage of deterioration, with a pungent smell of amines. Tentatively, it can be suggested that these corrosive sediments may have been formed by the burial of organic debris by submarine slides or turbidity currents.

The literature has indicated that an increased activity of SRBs results from increasing hydrostatic pressure [10]. The limited exposure tests reported here gave a maximum increase in the corrosion rate of about 100% at 30 atm as compared to 1 atm pressure. However, as the determination of free corrosion rates show a scatter, this value should be regarded as very tentative.

Conclusions

In-situ corrosion tests have been performed on superficial sediments (down to 5 m below the sea bottom) at depths down to 500 m on the Norwegian Continental Shelf. Corrosion studies have also been performed on soil samples collected from the seabed. Based on these investigations the following results are pertinent:

1. The corrosivity of the superficial sediments on the Norwegian Continental Shelf are generally low, being in the range 1 to 50 μm/year. For most sediments which have a glacial or glaciomarine origin and a low organic content, the corrosion rate will be in the range 1 to 20 μm/year.

2. In some cases the sediments contain very corrosive organic material and here the average corrosion rate can exceed 50 μm/year and may be as high as 360 μm/year. Such samples can be identified by their pungent odor.

3. These superficial sediments are mainly anaerobic. The corrosion rate of steel has been considered to be mainly determined by the sulfide electrolyte system created by the microbiological activity in the sediment. However, the SRB activity as determined by bacterial growth tests and counts can vary from 0 to 107 counts per gram and is thus not directly correlated to the corrosion rate.

4. As described in the paper, a good estimate of the corrosivity of the marine sediments can be given directly based on the soil description and soil properties. If a more exact corrosion evaluation is required, polarization studies should be performed on undisturbed core samples.

5. The polarization studies performed with steel in undisturbed samples produced anodic Tafel gradients in the range 30 to 55 mV/decade and cathodic Tafel gradients in the range 65 to 140 mV/decade. The corrosion rates determined by the cathodic Tafel gradient intercept method will generally be about 2 to 3 times higher than that determined by weight loss (one year's exposure). However, for sediments with a high in-situ content of sulfides, the laboratory evaluation can be erroneous due the secondary change in the chemistry of the sulfides as a result of pressure release and/or oxidation during sampling and transport.

6. Tentatively the CP maintenance current requirements in the majority of clay sediments on the deeper parts of the Norwegian Continental Shelf would seem to be in the range 10 to 15 mA/m^2.

Acknowledgment

Some of the data presented are from the Veritec-Joint Industry R&D project: "Corrosion Control in Seawater at Depths Down to 500 m." The authors wish to thank the project participants BP Norge A/S, Det norske Veritas/Veritec A/S, Elf Aquitaine Norge A/S, Esso Exploration and Production Norway Inc., Mobil Exploration Norway Inc., Norsk Hydro a.s, A/S Norsk

Shell, The Norwegian Geotechnical Institute, The Norwegian Petroleum Directorate, MARIN-TEK A/S, A/S Skarpenord and Unionoil Norge A/S for allowing us to utilize these data in the paper.

References

[1] Reinhart, F. M., "Corrosion of Metals and Alloys in the Deep Ocean (final)," Civil Engineering Laboratory, TR-834, National Technical Information Seminar 265, 1976.

[2] Di Gregorio, J. S. and Fraser, J. P., "Corrosion Tests on the Gulf Floor, in *Corrosion in Natural Environments, ASTM STP 558*, 1974, pp. 185-208.

[3] Ferguson and Wood, E. J., *Marine Microbial Ecology, Modern Biological Studies*, Chapman and Hall, Ltd. 1965, p. 243.

[4] King, R. A., "Prediction of Corrosiveness of Sea Bed Sediments," *Materials Performance*, Vol. 19, No. 1, 1980, pp. 39-49.

[5] Barker and Jørgensen, B., "Mineralization of Organic Matter in the Sea Bed—the Rate of Sulphate Reduction," *Nature*, Vol. 296, April 1982, pp. 643-645.

[6] Iversion, W. P., "An Overview of the Anaerobic Corrosion of Underground Metallic Structures. Evidence for a New Mechanism," *Underground Corrosion, ASTM STP 741*, E. Escalante, Ed., 1979, pp. 33-52.

[7] Miller, J. D. A. and Tiller, A. K., *Microbial Corrosion of Buried and Immersed Metal in Microbial Aspects of Metallurgy*, J. D. A. Miller, Ed., *Medical and Technical Publishers*, 1971, pp. 61-105.

[8] Morris, D. R. et al., "The Corrosion of Steel by Aqueous Solutions of Hydrogen Sulfide," *Journal of the Electrochemical Society*, Vol. 127, No. 6, 1980, pp. 1228-1235.

[9] Ramm, A. E. and Bella, D. A., "Sulfide Production in Anaerobic Microcosms," *Limnology and Oceanography*, Vol. 19, No. 1, 1974, pp. 110-118.

[10] Willingham, C. A. and Quinby, H. L., "Effects of Hydrostatic Pressure on Anaerobic Corrosion of Various Metals and Alloys by Sulphate Reducing Marine Bacteria," *Developments in Industrial Microbiology*, Vol. 12, 1970, pp. 278-284.

[11] Bolmer, P. W., "Polarization of Iron in H_2S-NaHS Buffers," *Corrosion*, Vol. 21, No. 3, 1965, pp. 69-75.

[12] Løken, T., "Geology of Superficial Sediments in the Northern North Sea," Publication No. 114, *Norwegian Geotechnical Institute*, Oslo, Norway, 1976, pp. 45-59.

[13] Gunnleiksrud, T., "Geologi og grunnforhold pi norsk sokkel—oversikt. Plattformer til havs, samvirke, jord og konstruksjon (in Norwegian)," *Kursdagene ved NTH*, 6-8 January 1986.

[14] Fischer, K. P., Cathodic Protection in Saline Mud Containing Sulphate Reducing Bacteria," *Materials Performance*, Vol. 20, No. 10, 1981, pp. 41-46.

[15] Fischer, K. P. and Bue, B., "Corrosion and Corrosivity of Steel in Norwegian Marine Sediments, in *Underground Corrosion, ASTM STP 741*, E. Escalante, Ed., 1979, pp. 24-32.

[16] Fischer, K. P. and Sydberger, T., "In-situ Testing of Cathodic Protection on the Northern Norwegian Continental Shelf," NACE, Corrosion/84, New Orleans, 2-6 April, paper No. 344 (1984), to be published, NACE, Houston, TX.

Thomas V. Edgar[1]

In-Service Corrosion of Galvanized Culvert Pipe

REFERENCE: Edgar, T. V., **"In-Service Corrosion of Galvanized Culvert Pipe,"** *Effects of Soil Characteristics on Corrosion, ASTM STP 1013,* V. Chaker and J. D. Palmer, Eds., American Society for Testing and Materials, Philadelphia, 1989, pp. 133–143.

ABSTRACT: The Wyoming Highway Department changed the criteria it was using to select corrosion protection levels for culvert pipe in 1981. A study was undertaken to compare the effect of these changes in selection criteria to corrosion developed on pipe in service. Twelve 35 to 40-year-old culvert pipes from four highway reconstruction sites were sampled to determine the maximum weight loss from the most corroded area of each pipe. In-situ and laboratory soil resistivities were determined in or adjacent to each pipe trench, along with soil pH and soluble salts including sulfates. Corrosive weight loss was shown to be related to the field resistivity but not to the other factors. This indicates that minimum resistivity and soil pH, the two most commonly used soil parameters for culvert pipe selection, may not be reliable corrosion predictors in practice.

KEY WORDS: corrosion, culverts, electrical resistivity, corrosion environments, field tests, soils, underground corrosion

Culverts are used to convey water through highway embankments when the estimated flow of water is not sufficient to require the use of a bridge. Being a part of the embankment, the culvert must be sufficiently strong to support the overlying embankment section, the roadway, and the expected traffic loads [1]. Failure of the culvert, due to any reason, can cause distress and possible failure in the road section. While it is possible that a culvert pipe could be structurally underdesigned initially, the most common reason for a culvert pipe to fail is due to a gradual weakening caused by corrosion. The rate of deterioration is a function of many factors, including properties of the pipe and its protective coatings, the nature of the soil, the chemicals in solution in the soil water, and the climatic and hydrologic regime in which the culvert exists.

The Wyoming State Highway Department changed its selection criteria for corrosion protection levels required for culvert pipe in 1981. The new criteria substantially raised the level of protection required for new pipe. In the field, however, it was recognized that often very serviceable 40-year-old zinc-coated pipe was being replaced with pipe having a much higher level of corrosion protection, usually either bituminous or polymeric-coated metal pipe or concrete pipe made with Type II or Type V cement.

A study was undertaken in 1985 to investigate the overall quality of pipe being removed at several Wyoming highway reconstruction sites and to qualitatively compare the actual corrosion on the old pipe to the corrosion protection specified for new pipe [2]. Based on the results of this study, the selection criteria were modified to more accurately represent the protection level required in the field.

[1]Assistant professor, Department of Civil Engineering, University of Wyoming, Laramie, WY 82071.

Project Description

Many states use the procedure developed in California to estimate the service life of galvanized steel culvert pipes [3]. This procedure uses the minimum resistivity of the soil and the pH of the soil and water to estimate the years to perforation of 16-gauge steel culvert. Other states, for example Utah [4], have determined that soluble salts including sulfates may also be useful predictors of corrosion rate in steel culverts. Experience has shown that the corrosion rates occurring in Wyoming tend to be significantly less than those estimated using these procedures.

The rate of corrosion is also influenced by the temperature and water content of the soil. Cooler temperatures reduce the chemical activity of the corrosion cell, and reduced water contents can increase the soil resistivity by several orders of magnitude greater than the value determined by the minimum resistivity test. Therefore, a corrosion rate criterion based on the worst case represented by the minimum resistivity test may not be appropriate in situations where a culvert may not be flowing 300 or more days per year.

Test Sites

This project combined field and laboratory studies to investigate the factors affecting culvert corrosion and the Wyoming Highway Department's corrosion protection criteria. Four highway reconstruction sites were investigated where high corrosion resistance pipe had been specified due to either low values of minimum resistivity or high values of soluble salts. Specifications for this pipe require polymeric precoated galvanized pipe with 10-mil coatings on both sides, asbestos-bonded, bituminous-coated galvanized steel pipe, reinforced concrete pipe with Type V cement, or vitrified clay pipe. The contractor can select the type of culvert used based on cost, site conditions, and previous experience.

These sites represented a range of climates and water conditions common in the southern half of the state. Average rainfall at the sites ranged from 250 mm at Baggs to 500 mm at Beaver Creek. Similarly, average annual temperatures ranged from 3 to 12°C. Three sites had elevations over 2200 m while the fourth had an elevation of 2000 m.

Field Procedures

The basic field procedure used consisted of four stages. After the pipe was excavated and removed, the field resistivity of the soil was determined along the centerline of the trench. A bag sample of representative soil was collected from around the pipe along with a sample for the determination of water content. Finally, a pipe specimen 0.6 m long was obtained from the pipe section experiencing the most severe corrosion. This procedure was not always possible since scheduling problems occasionally made it necessary to obtain samples of the pipe after it had been removed and the new pipe installed.

Soil Resistivity—Soil resistivity is a measure of the ability of the soil to allow an electrical current to flow. Measurement of the field resistivity is the most direct indicator of the local electrical properties. Soils which have a low resistivity allow a large current to pass through the soil, causing a large metal loss. Conversely, soils with a high resistivity permit only a small current with an associated low metal loss. The testing followed ASTM Method for Field Measurement of Soil Resistivity Using the Wenner Four-Electrode Method (G 57–78).

Resistivity can vary significantly with depth. In general, surface readings along the bottom of the trench at the invert location would have been preferred. This is not possible using in-situ techniques since the spacing of the probe pins must be large relative to the depth the pins are driven. The value of resistivity reported in this study was the average of two 4-ft spacing readings and one 8-ft spacing reading usually taken in the excavated culvert trench. The two 4-ft spacings overlapped the 8-ft spacing. It was believed that this average would compensate for any local variations and also be weighted to the shallower readings.

In those locations where the new pipe had already been installed, resistivity readings were obtained parallel to the road and along the base of the embankment at the ditch line. Comparison readings indicated that this technique reasonably approximated the readings which were determined in the pipe trench.

Soil Moisture—One factor influencing the resistivity values of a soil is its moisture content. Soils with high moisture content generally have lower resistivities. Soil samples for determining water content were taken at each culvert excavated while the investigation team was there. However, several pipes were excavated and replaced while the team was not present. In these cases, since the soil had an opportunity to dry, water content samples were not obtained. An insufficient number of water content samples were obtained to be significant, therefore no conclusions have been drawn from that data.

Bag Samples—Bag samples were obtained from each site. The sample weights ranged from 2 to 8 kg. They were usually taken from the center of the trench and adjacent to the pipe sample. In several cases in which the pipe was previously excavated, the sample was taken from the soil which stuck to the side of the pipe at the pipe sample location or close to it. The soil chemistry tests were performed on these samples.

Pipe Samples—In general, it was attempted to obtain pipe sections which were undamaged during excavation and removal. Pipe sections 0.6-m long were cut from the pipe using an acetylene torch. There appears to be some damage to the zinc coating adjacent to the cut due to the flame, but the pipe sections obtained were large enough that undamaged specimens could be cut away from the affected areas.

Inspection of the excavated culverts showed that the region of heaviest corrosion was typically in the center portions of the pipe and that damage decreased toward the ends. Most culverts were constructed from two pipe sections and joined together with a band. It was possible to cut the bolts on the band and make one circumferential cut around the pipe to provide both the band and a 2-ft-long pipe section for testing. This provided the required sample with a minimum of cutting.

Laboratory Procedures

The field sampling was augmented by laboratory testing of the soil and pipe samples. The pipe specimens were cut from the pipe sample, cleaned, weighed, and compared to the average weight of several uncorroded specimens of the same size and gauge. The soil testing was performed using the routine, standardized testing procedures developed by the Wyoming Highway Department. The soil tests performed were the minimum resistivity test, soil pH, total soluble salts, and sulfates.

Pipe Specimens—Test specimens were cut in the most corroded areas of the pipe sections using a 100-mm-diameter saw drill. The pilot holes were carefully centered on the top ridges of the pipe corrugates to insure that the samples would be uniform. Control specimens were prepared using the same technique but were taken from uncorroded sections of the pipe. The reproducibility of this technique was confirmed by comparing the weights of the control specimens. They usually weighed within 1 g of each other, producing a variation of less than 0.5%.

After the specimens were cut, they were stamped with identifying numbers. They were then soaked in a solvent to remove the cutting oil, attached soil, and some of the corrosion from the surface. Each specimen was hand brushed with a wire brush and then finally cleaned with a wire brush mounted on a bench grinder. The weights of each specimen were then recorded. The weight loss on the culvert pipe specimens was determined by subtracting the actual weight of the specimens from the average weight of the control specimens of the same gauge.

Attempts were initially made to chemically clean the specimens. It was determined that the amount of soaking required to clean the surface was also sufficient to cause significant loss of uncorroded steel. The brushing technique may have left a small amount of corroded metal on the specimen or removed a little more of the more deeply pitted sections of the otherwise intact steel, but the remainder of the specimen was that portion which provided the basic structural strength of the culvert.

Minimum Resistivity—The minimum resistivity test determines the lowest possible value of resistivity that a soil sample may have. It represents the worst possible resistivity condition of the soil and may be much less than the actual resistivity in the field. It is performed in a manner similar to the field resistivity in that a current is passed through soil in a box of standard dimensions and the voltage is determined to find the soil resistivity. A sample is prepared by sieving the soil through a No. 8 sieve and adding water to the finer fraction to obtain a water content of about 12%. The soil is then packed into the box by hand and the resistivity of this mix obtained. The water content is then raised to about 16% in two increments, and the test is performed at each water content. If the resistivity did not reach a minimum at the second point, the water content is raised in successive increments of 4%, and the resistivity measures are made at each increment until the minimum value is found. Note that the soil may be well past the point of initial saturation before the minimum resistivity is found.

Soil pH—The pH of the soil is determined using a special pH electrode cup for soil. The soil is mixed at a one-to-one ratio by weight with distilled water. This mix is placed in the electrode cup and the value of pH read directly off the meter.

Soluble Salts—The percentage of the solid weight represented by soluble salts is determined using saturation techniques. The salts typically reflect the ratio of the amount of rainfall to the amount of soil surface evapotranspiration. If the evapotranspiration rate is high, water with the soluble salts will flow to the surface from below due to capillary action. When the water evaporates or transpires, the salts are left in the upper crust of the soil. The value of the percent salts is extremely variable in lateral extent and, in arid regions, decreases rapidly with depth in the soil.

Typically, the electrical resistivity of the soil is a function of the water content and the amount of soluble salts in the water solution. The greater the salt content, usually the lower the resistivity. This is not a one-to-one ratio, however, since the different ions available have different weights associated with one charge. Similarly, high water contents typically produce low-resistivity values since there are both large areas of water for the current to flow through and more complete hydration of the ions.

A wide variety of salts typically exists in the soil solution [5]. The most common cations in a soil solution are potassium, sodium, magnesium, and calcium. Common anions include carbonates, sulfates, chlorides, and oxides. Each of theses ions has its own reactivity with metal surfaces. Two soils having the same minimum resistivity may react differently depending on the specific ions available in each. Thus, the amount of total salts available may not be indicative of the corrosivity of the solution if certain combinations of compounds may form. For example, calcium and magnesium tend to form insoluble oxide and carbonate precipitates in basic environments which can create a protective layer over the metal surface and reduce the corrosion.

Sulfates—The percent sulfates is a measure of the amount of sulfate in the total dry weight of the soil. Being one of the soluble salts, its value must be lower than the percent soluble salts described above. The reason why the sulfates were tested is because it provides a measure of the potential reaction of the soil to portland cements, which are high in tricalcium aluminate. Thus, if the contractor uses concrete pipe for the culverts, a sulfate level greater than 0.1% indicates that a sulfate-resistant Type II or Type V pipe is required.

A possible secondary source of corrosion could be caused by sulfate-reducing or sulfur-oxidizing bacteria. Two genera of sulfate-reducing bacteria are significant, the Desulfovibrio and the Desulfotomaculum. These reduce the sulfate in anaerobic and moderate pH conditions together with hydrogen or organic matter to form ferrous sulfide and ferrous hydroxide [6, 7].

Sulfuric acid is formed by the oxidation of sulfur though the metabolism of microbes of the genus Thiobacillus [5, 7, 8]. Sulfuric acid concentrations up to 12% have been produced by

Thiobacillus Thio-oxidans. Low pH's, in the order of 2 to 4, are required for this microbe or one of several others to develop.

General Observations Concerning Corrosion on Culvert Pipe

All pipes were installed between 1946 and 1948 and were excavated in 1985. Most of the pipe inspected was in good to very good condition, including several sections which could have gone back into the embankment. The manufacturer's stamps could often be seen and read on the pipe. The corrosion was usually in local areas on the pipe, indicative of the formation of small corrosion cells. This is significant since even in the worst individual specimen, Riverside 5, the majority of the pipe was structurally sound with only small local weaknesses. Since the water in the culvert is not under high pressure as in a pipeline, the embankment would usually still be safe even if the culvert pipe developed pinhole leaks. Continued corrosion, however, could finally reduce the overall thickness of the pipe sufficiently to affect the structural integrity of the embankment.

Usually the most corroded area was in the center of the culvert in what is considered to be the most anodic section in an oxygen concentration cell under a roadway [9]. Commonly, a joint was located under the centerline of the road at which two sections of pipe were banded together. It was observed in most culverts that the band experienced the worst corrosion on the entire culvert. In several cases, the band had corroded through and thus provided neither support nor alignment to the pipe. This former function is not necessary, of course, and the latter is not critical once the embankment is in place. The section of pipe covered by the band was entirely free of corrosion, even when there was otherwise severe corrosion on the pipe immediately adjacent. Thus, this section provided full strength under the corroded band.

One other observation is significant. In most studies, it has been indicated that the worst corrosion typically occurs at the invert, or bottom, of the pipe. There are several possible reasons why this occurs. Standing water in and around the invert would reduce the oxygen supply to this region, making it anodic to the crown, or top, of the pipe. This was not consistently observed in the study. Instead, the side walls often showed as much corrosion as the invert. This would tend to indicate that the water content of the soils was probably uniform around the pipe, not allowing any excessive oxygen deprivation along the bottom of the pipe.

Pipe Gages

Pipe gages presented some problems in this analysis. Since weight loss was determined by subtracting the weight of the corroded 100-mm-diameter sample from the uncorroded control specimen, small variations in control specimen weights had a significant effect on the measured weight loss. Five distinct pipe thicknesses were encountered in the 13 specimens. Gauges 12 and 16 were similar to their specifications, as shown in Table 1. Assuming the rest to be 14 gauge, several corroded specimens weighed more than the control specimens used for comparison.

Further investigation of specimen thicknesses and weights indicated that 14 gauge appeared to have three different thicknesses. The thickest was designated 13 gage, even though the thickness was 0.15 mm less than the standard gage thickness. The other two sizes may have been due to different roller spacings or to uneven zinc thicknesses caused by irregular dipping. The lighter specimen was designated 14 gage and the heavier was designated 14+ gage. The final weights used are given in Table 1.

The 14+ gage was not consistent along the length of the pipe and the determined weight loss seemed much higher than the appearance would indicate. Pipe having this gage appeared at three sites and may have been a viable thickness in the middle 1940s to 1950s. This variation will be mentioned later as a possible source of error.

TABLE 1—*Pipe specimens and gages.*

Nominal Gage	Standard Thickness, mm	Measured Thickness, mm	Average Weight, g	No. of Samples Tested
16	1.63	1.68	100.66	2
14	2.01	1.96	119.85	2
14+	2.01	2.06	127.51	2
13	2.39	2.24	136.56	2
12	2.77	2.77	168.13	3

Analysis of Data

Many states use some combination of minimum resistivity, pH, and soluble salts as a basis for selecting the required pipe. The data collected in this project provide an opportunity to compare these quantities, along with the measured field resistivity, to the actual amount of metal lost on several culvert pipes.

Table 2 presents a summary of the field and laboratory data. Several sites had readings taken both upstream and downstream from the centerline of the roadway. At Riverside 1, Specimen 8, the embankment was reconstructed and readings were taken at the base of the embankment on both the upstream and downstream sides of the road. The readings at Riverside 4, Specimen 12, were both taken in the culvert trench. The nature of the soil changed right at the centerline, so both those readings were taken in the trench, upstream and downstream of the center of the roadway. Specimen 16 is similar to Specimen 12 except it is taken from the band and not the pipe.

Figure 1 compares the weight of metal lost per unit area to the field resistivity. Considerable data scatter is shown over the resistivity range. The six worst outrider points, having weight losses of 10 g/dm² or more and resistivities greater than 5000 ohm-cm, each represent questionable or otherwise problematic data. The top four points, showing weight losses of 59.8 and 37.33 g/dm², are from Riverside 4. Two significant factors existed at this location. First, a highly organic topsoil was used as the fill material. Secondly, this culvert was located one-half on a cut and one-half on a fill of a different material. This created a strong local corrosion cell. Since the resistivity values are low, this would have increased the rate of corrosion at the interface.

The other two questionable points, representing 23.2 and 14.4 g/dm² loss, were both from pipes having the 14+ gauge. If the weights of the pipe sections are subtracted from the weight of the standard 14-gauge controls, the weight losses become 13.6 and 4.7 g/dm², respectively. This would place the points back into the region of the other points now below them.

Figure 2 results if the points from Riverside 4 and the points having the 14+ gage pipe are removed from Fig. 1. Figure 2 shows much less scatter between weight loss and field resistivity. A best-fit line for this data is

$$WL = 880 \, (FR)^{-0.565}$$

where WL is the weight loss (g/dm²) and FR is the field resistivity (ohm-cm). This relationship has a correlation coefficient of -0.831. Also shown is the line at two standard deviations above the average line. This second line could be used as an estimate for the maximum probable weight loss relative to a given field resistivity. This line has the general equation

$$WL_{2s} = 1760 \, (FR)^{-0.565}$$

TABLE 2—*Pipe Specimen and Site Characteristics*

Specimen No.	Location	Gage	Pipe Diameter, mm	Test Location, T/E[a]	Specimen Weight, g	Weight Loss, g	Weight Loss/Area, g/dm²	Field Resistivity ohm-cm	Min Resistivity ohm-cm	pH	Soluble Salts, %	Sulfates, %
1	Beaver Creek	12	915	T	159.14	8.99	11.2	4 200	650	7.45	0.20	0.0047
2[a]	Beaver Creek	16	230	E	92.14	8.52	10.6	3 200	1850	8.01	0.04	0.0021
2[b]	Beaver Creek	16	230	E	95.51	5.15	6.4	3 200	1850	8.01	0.04	0.0021
3	Beaver Creek	14+	610	E	123.50	4.01	5.1	17 000	1500	7.54	0.09	0.0027
4	Beaver Creek	14	610	T	103.04	16.81	21.0	1 900	705	7.07	0.22	0.0080
5	Danial Junction	16	760	E	96.28	4.38	5.5	7 800	1400	8.22	0.11	0.0112
6	Danial Junction	14+	610	T	108.94	18.57	23.2	5 500	1770	8.02	0.10	0.0041
7	Danial Junction	16	460	E	93.78	6.88	8.6	1 700	625	8.08	0.13	0.0249
8	Riverside 1	12	915	E	165.76	2.37	3.0	up25 800[b] dn11 200	1550	7.44	0.11	0.0028
10	Riverside 2	14+	610	T	115.97	11.54	14.4	11 800	2100	6.13	0.09	0.0033
11	Riverside 3	14	610	T	116.25	3.60	4.5	6 100	2100	6.28	0.12	0.0041
12	Riverside 4 (Pipe Section)	13	610	T	106.84	29.83	37.3	up11 000[c] dn 5 700	3300 1900	7.74 7.52	0.06 0.10	0.0014 0.0023
16	Riverside 4S (Band Section)	16	610	T	52.86	47.80	59.8	up11 000[c] dn 5 700	3300 1900	7.74 7.52	0.06 0.10	0.0014 0.0023
13	Riverside 5	12	915	E	159.25	8.88	11.1	2 800	1950	7.22	0.09	0.0026
14	Baggs	14	760	E	111.72	8.13	10.2	2 800	2100	6.77	0.08	0.0026

[a]Field Resistivity measurements were taken in the pipe trench (T), or adjacent to the embankment (E), perpendicular to the pipe.
[b]Field Resistivity measurements were taken on the upstream (up) and downstream (dn) sides of the embankment.
[c]Field Resistivity measurements were taken on the upstream (up) and downstream (dn) of the lane centerline in the trench.

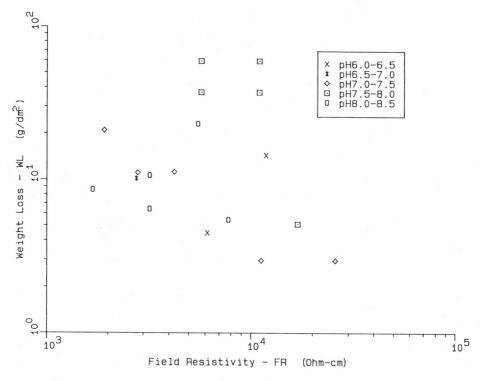

FIG. 1—*Weight loss for all specimens as a function of field resistivity and soil pH.*

An upper bound estimate of this type could be incorporated into a design method for culvert pipe selection based on a strength reduction as a function of weight loss.

Figure 2 also shows that, for the group of soils considered, the effect of pH is negligible. It has been indicated [10] that when the pH of the soil is between 6.0 and 9.0, it has little or no effect on the corrosion. Therefore, there appears to be little reason to include pH in a design method for values within this range.

Figure 3 demonstrates the relationship between weight loss and minimum resistivity for the same points shown in Fig. 2. The expression for this line is

$$WL = 346 \ (MR)^{-0.537}$$

where MR is the minimum resistivity in ohm-cm. The scatter shown here is less predictable than that of Fig. 2, with a correlation coefficient of -0.422. Thus, it appears that minimum resistivity, or even minimum resistivity coupled with pH, cannot reasonably predict the corrosive weight loss for these specimens. Two situations may account for this observation. Soils in which the minimum resistivity is reached at water contents well above saturation cannot reach that water content in the field. Thus, the potential described by the minimum resistivity can never be attained. Secondly, most soils in this study were at water contents well below saturation and typically would remain so during all or most of the year. Only if the soil were more completely saturated would the field resistivity approach the minimum resistivity. In either case, the field resistivity more accurately reflects the in-situ conditions causing corrosion than does the minimum resistivity.

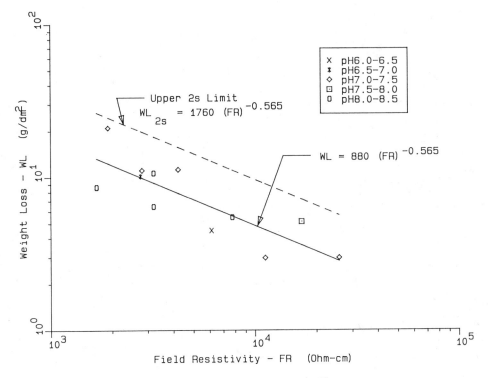

FIG. 2—*Weight loss of selected specimens as a function of field resistivity and soil pH.*

Another factor commonly used in design methods is the percent soluble salts. Figure 4 presents the weight loss versus the percent soluble salts. There does not appear to be any significant relationship between these two values. The data indicate that a reasonable correlation exists between the percent soluble salts and the minimum resistivity. As the percentage of ionic salts increases in the water, the conductance of the soil also increases. Thus it is redundant to check for both minimum resistivity and soluble salts.

Conversely, the individual salts in the preparation solution were not determined except for sulfates. The percent sulfates in the samples were low and did not have a predictable effect on the corrosion of these samples. While specific ions may have influenced the corrosion rate, the data tested was insufficient to determine this effect.

Conclusions

It must be emphasized that only a limited number of samples and sites were tested. The results, therefore, are not conclusive. There are definite trends indicated in the data, however, which could lead to a more rational selection criterion for culvert pipe in Wyoming.

Based on visual inspection and laboratory testing of the most highly corroded specimens of culvert pipe collected on this project, it appears that most culvert pipe should have been replaced with regular galvanized culvert pipe rather than the coated pipe specified.

There is a good relationship between metal lost and field resistivity. It is likely that a selection criteria for corrosion-resistant culvert pipe could be developed using field resistivity as either the primary or the only guideline. This would be very applicable in reconstruction jobs in which the

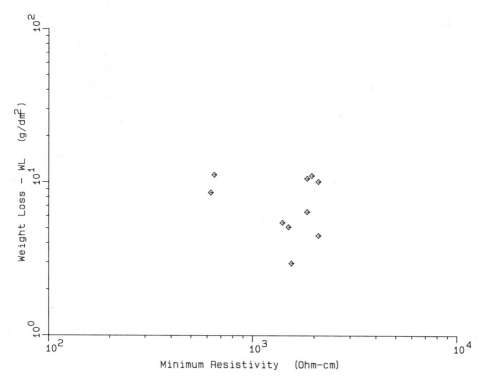

FIG. 3—*Weight loss of selected specimens as a function of minimum resistivity.*

embankment has established an equilibrium water content. Although no long-term testing was performed, it is likely that seasonal fluctuations would have little effect on field resistivities measured alongside an embankment parallel to the road centerline or actually inside the embankment.

Field Conclusions

This project provided an excellent opportunity to observe the effects of a 35 to 40-year-old construction procedure. The most obvious defect in practice was the poor backfill material used. Often it was the local topsoil. There were few trenches which had been filled with imported material, even when it would have been appropriate. If questionable material is shown during excavation in the trench, a clean, coarse, cohesionless backfill material should be used.

Even more important is the case in which two different soils interface with the culvert. The local cell reaction can be severe, causing corrosion rates two or three times higher than would have occurred in either soil. In this situation, it is mandatory that a controlled backfill of clean, coarse, cohesionless backfill be used. The reason for using coarse backfill is that the natural resistivity is lower when the soil is drained. Secondly, better oxygen penetration, both in damp and saturated conditions, reduces the possibility of oxygen concentration cells developing.

Acknowledgments

The Wyoming Highway Department and the University of Wyoming provided the support for the work presented. Sterling Roberts and Jeffrey Groom aided the author in this project. The

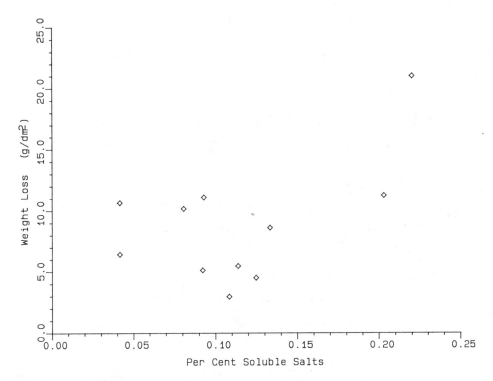

FIG. 4—*Weight loss of selected specimens as a function of soluble salts.*

opinions expressed herein are those of the author and do not necessarily reflect those of the sponsoring agencies.

References

[1] "Durability of Drainage Pipe," National Cooperative Highway Research Program, Synthesis of Highway Practice 50, Transportation Research Board, National Research Council, Washington, DC, 1978.

[2] Edgar, T. V., "Analysis of the Wyoming Highway Department Corrosion Resistance Criteria for the Selection of Culvert Pipe," Department of Civil Engineering, University of Wyoming, Laramie, WY, January 1986.

[3] Beaton, J. L., and Stratfull, R. F., *Proceedings, Highway Research Board,* Vol. 41, National Research Council, Washington, DC, 1962, pp. 255-272.

[4] Welch, B. H., Transportation Research Record, No. 604, National Research Council, Transportation Research Board, Washington, DC, 1976, pp. 20-24.

[5] Shreir, L. L., ed., *Corrosion, Vol. 1: Metal/Environmental Reactions,* Newnes-Butterworths, London, 1976, pp. 2:61-82.

[6] Uhlig, H. H. and Revie, R. W., *Corrosion and Corrosion Control,* 3rd ed., John Wiley and Sons, New York, 1985.

[7] Hamilton, W. A., "Microbial Corrosion," The Metals Society, London, 1983, pp. 1-5.

[8] Bos, P. and Kuenen, J. G., Microbial Corrosion, The Metals Society, London, 1983, pp. 18-27.

[9] Peabody, A. W., "Control of Pipeline Corrosion," National Association of Corrosion Engineers, Houston, TX, 1967, p. 7.

[10] Haviland, J. E., Bellair, P. J., and Morrell, V., "Durability of Corrugated Metal Culverts," Highway Research Record No. 242, National Research Council, Highway Research Board, Washington, DC, 1968, pp. 41-66.

Gardner Haynes,[1] Gregory Hessler,[2] Reiner Gerdes,[3] Kenneth Bow,[4] and Robert Baboian[1]

A Method for Corrosion Testing of Cable-Shielding Materials in Soils

REFERENCE: Haynes, G., Hessler, G., Gerdes, R., Bow, K., and Baboian, R., "A Method for Corrosion Testing of Cable-Shielding Materials in Soils," *Effects of Soil Characteristics on Corrosion, ASTM STP 1013,* V. Chaker and J. D. Palmer, Eds., American Society for Testing and Materials, Philadelphia, 1989, pp. 144–155.

ABSTRACT: A method for testing shielding materials for corrosion resistance is presented. Techniques for preparing specimens and evaluating the effects of alternating current on corrosion are described. Analytic methods are discussed for determining soil constituents which impact corrosion. The results of a six-year burial program for cable-shielding materials are listed.

KEY WORDS: soil corrosion, characterization of soils, cable shielding, corrosivity of soils, a-c corrosion, soil water extracts, copper clad steel, ASP, CASP, CACSP, aluminum clad steel, corrosion testing in soils

In order to function properly, a cable sheath must provide mechanical protection to the cable core during and after installation, prevent ingress of moisture into the cable core (in certain designs), and provide electrical conductivity for the life of the cable. Thus, mechanical strength, electrical conductivity, and corrosion resistance are important criteria for cable-shielding materials. The operational characteristics of shield designs have been compared by a number of investigators [1–6].

The impact of corrosion on shielding materials has also been studied [7–9]. The National Bureau of Standards (NBS) and the Rural Electrification Administration (REA) have investigated the corrosion resistance of shielding materials through extensive burial tests [10–12]. These studies included over 100 different shielding materials and employed electrically isolated lengths of cable which had damage sites (windows and rings) in the jacket as well as similar lengths of cable which were galvanically coupled to copper strips. The tests were conducted at established NBS test sites whose soil characteristics have been documented [10].

The comparative behavior of idle cable versus cable-carrying alternating current, as in actual service, was not evaluated in the NBS-REA tests.

The present study was a cooperative effort by REA, Horry Telephone Cooperative, The Dow Chemical Co., Contel Corp., and Texas Instruments Inc. to determine the corrosion behavior and the associated effects of alternating current which is present in service on commonly used shielding materials. This paper describes the test program, presents analytical methods for determining soil constituents, and lists the results of the burial tests.

[1]Member, Group Technical Staff, and head, respectively, Electrochemical and Corrosion Laboratory, Texas Instruments Inc., 34 Forest St., Mail Station 10/13, Attleboro, MA 02703.
[2]Communications specialist, REA-Outside Plant Branch, Telecommunications Engineering & Standards Division, Washington, DC 20250.
[3]Director, Contel Laboratories, Contel Corp., 270 Scientific Dr., Suite 10, Norcross, GA 30092.
[4]Senior associate development scientist, The Dow Chemical Co., P.O. Box 515, Bldg. B, Granville, OH 43023.

Test Program

The commercially available shielding materials included in the test program are listed in Table 1. The shielding materials were fabricated into 25 pairs of polyethylene-jacketed telephone cable according to REA specifications. Continuous 500-ft lengths of cable with damage sites at 30-ft intervals as well as individual two and one-half foot lengths of cable with damage sites were prepared by the shielding manufacturers. The damage site pattern (Fig. 1) was intended to simulate possible construction, lightning, or rodent damage to the cable's outer jacket. The circumferential rings and windows were made by cutting the polyethylene jacket to within 0.010 in. of the shield with a single-edge razor blade in a depth-controlled fixture. The polyethylene jacket was then peeled away, taking care not to damage the shield. The window was designed to promote differential aeration and give an indication of the tendency of any spot damage to the cable to result in corrosion around the circumference of the shield. The holes were made by drilling to within 0.010 in. of the shield with a collared drill and removing the remaining jacket by hand with an end mill. The ends of the individual lengths of cable were sealed with vinyl tape and wax. This type specimen allows comparison of the amount of corrosion at the windows and holes to that occurring where the complete circumference is exposed.

In March of 1980 the 500-ft lengths of cable were buried in a test site under the jurisdiction of Horry Telephone Cooperative in Conway, SC. A trenching machine was used to dig a 4-in.-

TABLE 1—*Shielding materials in test program.*[a]

ASP	Bare 8-mil aluminum/bare 6-mil, tin-plated steel
CASP	Coated 8-mil aluminum/bare 6-mil, tin-plated steel
CACSP	Plastic-coated, 8-mil aluminum/plastic-coated, 6-mil, tin-plated steel
Al/ST/Al	Aluminum, clad steel, clad aluminum—0.008/0.003/0.005
CAl/ST/CAl	Coated aluminum, clad steel, clad-coated aluminum—0.008/0.003/0.005
Cu/ST/Cu	Copper, clad steel, clad copper—0.0004/0.002/0.0036

[a] Sheath interfaces were fully flooded with atactic polypropylene-based floodant, and the cable core was filled with petrolatum-based filler.

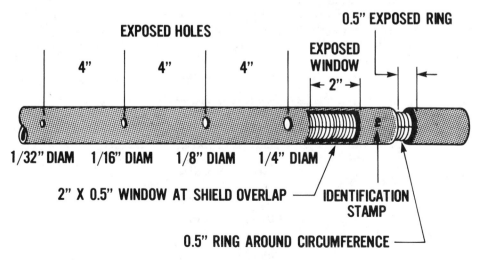

FIG. 1—*Pattern of damage sites for isolated lengths and continuous cable.*

wide, 3-ft-deep trench, and the cables were laid in by hand. Markers were placed on the nearby road to facilitate location of the test area for retrieval. The cables were electrically connected (dynamic) in series. The 3000-ft-long test cable was connected in parallel to an existing cable which had a high level of induced a-c shield current. Soil samples were obtained at the bottom of the trench at the midpoint of each test cable. The soil samples were sealed in locking plastic bags for transportation back to the laboratory. Seepage of water into the trench indicated that the cables, at the time of installation, were below the water table. The electrically isolated (static) specimens were buried at the same depth as the cable six feet away from each area with damage sites so that a pair of dynamic and static specimens could be easily obtained at each retrieval. The test program called for retrieval of a control cable (static) and a section of test cable (dynamic) of each type of shielding at intervals over the course of several years. Electrical continuity was maintained by splicing the cables each time a specimen was removed.

Soil Characterization

The pH and resistivity of the soils were measured within 24 h after collecting the samples. The pH was measured according to ASTM Test Method for pH of Soil for Use in Corrosion Testing (G 51-77). The values were measured in the laboratory using an Orion Model 801 pH meter. Resistivity values were measured in the laboratory according to ASTM Method for Field Measurement of Soil Resistivity Using the Wenner Four-Electrode Method (G 57-78). A Miller soil box and a Nilsson Model 400 soil-resistance meter were used to measure the "as-received" and "saturated" soil resistivity.

Preparation of soil water extracts was conducted using the techniques established by NBS-REA in previous studies [13]. A suspension of soil and distilled water in the ratio of 1:5 was shaken mechanically on a ball mill at intervals (8 h on, 16 h off) for a period of 72 h. The extract was decanted and filtered through 45 μm filters prior to analysis.

Qualitative analysis of the extracts was conducted using d-c plasma emission spectroscopy and a series of anionic specific methods. The primary chemical constituents were then quantitatively determined by the methods listed in Table 2. EPA water quality standards were used as an external check on the methods employed.

Cation (Na^+, K^+, Ca^{++}) analysis was readily accomplished with an atomic absorption spectrometer. Anions required more specific techniques. Carbonate and bicarbonate were determined by titration with 0.05 N sulfuric acid with a pH meter used to determine the endpoint. A pH of 8.2 was used as the endpoint for carbonate determination, while a pH of 4.5 was used as the endpoint for bicarbonate [14]. Chloride was measured with a Beckman ion specific elec-

TABLE 2—*Analytic techniques for soil water extracts.*

Species	Technique Employed
Na^+	Atomic absorption spectroscopy
Ca^{++}	Atomic absorption spectroscopy
K^+	Atomic absorption spectroscopy
Mg^{++}	Atomic absorption spectroscopy
Cl^-	Ion selective electrode
NO_3^-	Ion selective electrode
$SO_4^=$	UV-visible spectroscopy
HCO_3^-	Titrimetry
$CO_2^=$	Titrimetry

trode and a Beckman select ion 5000 analyzer. The reference electrode was a double junction reference electrode: outer compartment 10% HNO_3 and inner compartment Orion 90-00-02 (matches characteristics of saturated calomel electrode). All samples and standards were treated with 3-mL concentrated HNO_3 and 25-mL isopropyl alcohol to maintain constant ionic strength and remove interferences. Sulfate determination was made using the turbidimetric method [12]. This technique involves precipitation of sulfate by barium chloride and spectrophotometric determination of sulfate based on barium sulfate absorption at 420 nm. Nitrate was measured with an Orion 93-07 nitrate electrode and a Beckman select ion 5000 analyzer. The outer compartment of the double junction reference electrode was filled with 2 mL of 2 M ammonium sulfate and 100 mL of distilled water, while the inner compartment contained Orion 90-00-02. Strong interference in nitrate determination exists if high concentrations of chloride, bicarbonate, or carbonate are present. For this reason samples were acidified to a pH of 4.5 and chloride precipitated with silver sulfate. Precipitated silver chloride was removed by vacuum filtration through a 0.45-μm filter.

Results

Soil Properties

The properties of the soils at the Conway test site are shown in Table 3. The resistivity values for the "as received" soil samples are listed in parenthesis and range from a low of 11 000 to a high of 110 000 ohm-cm. The "saturated" resistivity values ranged from 7000 to 34 000 ohm-cm. As stated in G 57, these values should be considered a worst-case situation.

The pH values indicate that the soils are nearly neutral. As indicated by the high resistivity values, very low concentrations of soluble salts were found in all of the soils. The nearly complete absence of aggressive anions such as chloride, which accelerate corrosion, should be noted.

The corrosivity of the soil at the Conway test site can be estimated by comparing its properties with those of established test sites. Comparison of the soil resistivity and pH listed in Table 3 for Conway with the data in Table 4 for the NBS-REA test sites for cable shielding materials shows that Conway is similar to, although less corrosive than, Hagerstown loam. Note that no chemical analysis was made for Hagerstown loam by NBS-REA due to the very low concentrations of soluble salts in this soil. The modern analytical techniques described in this paper make chemical analysis of high resistivity soils (such as Hagerstown loam) feasible.

TABLE 3—*Properties of soils at CONWAY, SC test site.*

Shield	Resistivity, ohm-cm		pH	Composition of Water Extract, ppm							
	R[1]	R[2]		Ca	Mg	Na	CO_3	HCO_3	SO_4	Cl	NO_3
CASP	9500	(16 000)	5.47	0.6	0.4	7.2	nil	0.18	3	0.1	3
ASP	34000	(110 000)	5.40	1.9	0.4	5.4	nil	0.11	6	0.1	2
CACSP	11000	(29 000)	5.81	5.9	0.6	3.6	nil	0.10	15	0.1	8
Al-ST-Al	7000	(11 000)	5.55	3.1	0.8	10.8	nil	0.26	3	0.1	20
Cu-ST-Cu	11000	(19 000)	5.45	2.5	0.6	6.3	nil	0.13	4	0.1	16
CAl-ST-CAl	8000	(13 000)	5.40	3.4	0.7	9.4	nil	0.13	5	1.0	21

[1]Soil saturated (H_2O).
[2]Soil, as received.

TABLE 4—Properties of soils at NBS-REA test sites.

Site	Soil	Location	Internal Drainage of Test Site	Resistivity, ohm-cm[a]	pH	TDS[b]	Composition of Water Extract, ppm							
							Ca	Mg	Na + K as Na	CO$_3$	HCO$_3$	SO$_4$	Cl	NO$_3$
A	Sagemoor sandy loam	Toppenish, WA	Good	400	8.8	7 080	108	23	1960	0.0	5002	216	330	6
B	Hagerstown	Loch Raven, MD	Good	5 200	5.8	...[c]	...	...	...	...	...	...	...	...
C	Clay	Cape May, NJ	Poor	300	4.0	14 640	540	754	2242	0.0	0.02	6768	3529	118
D	Lakewood sand	Wildwood, NJ	Good	30 000	7.3	...[c]	...	...	...	...	...	...	...	...
E	Coastal sand	Wildwood, NJ	Poor	55	7.1	11 020	302	329	3230	0.0	55	1133	5765	31
G	Tital marsh	Patuxent, MD	Poor	300	7.1	11 580	140	165	2392	0.0	0.0	1709	3259	37

[a] Resistivity determinations made at the test site with Shepard canes.
[b] TDS = total dissolved solids—residue dried at 105°C.
[c] Analyses not made for soils at Sites B and D because of the very low concentration of soluble salts.

An indication of the corrosivity of the soils at Conway can be obtained by comparing the resistivities in Table 3 to those listed in Table 5. The indications of soil corrosivity as a function of resistivity given in Table 5 have been published by the National Association of Corrosion Engineers as a part of their basic corrosion course. According to this guide, the Conway soil should be considered less than mildly corrosive. However, other site factors such as drainage, aeration, etc. must be taken into consideration.

Shielding Specimens

A static and dynamic specimen of each type of cable was retrieved after 0.5, 1, 2.5, and 4-year exposure periods. Duplicate specimens were retrieved after six years of exposure. The amount of a-c current flowing in the cables after the six-year test period is listed in Table 6. After retrieval, the polyethylene jackets were removed from the specimens by personnel at REA and the flooding compound was dissolved in kerosene, taking care not to remove corrosion products so that the extent of corrosion could be evaluated. The specimens were then rated by a panel using the rating system in Table 7. The rating system had a scale of 0 to 10 with 10 being no indication of corrosion and 0 being electrical discontinuity due to corrosion. The panel consisted of representatives from REA, Contel Corp., The Dow Chemical Co., and Texas Instruments Inc.

Results from this test program are listed in Tables 8 through 12. Since the ratings were consensus opinions of a panel, explanatory notes were required and are given in Table 13. The intent of this paper is to record these results in the literature without bias, and, therefore, no discussion or interpretation of these results is included.

Summary

A field testing technique for corrosion resistance of shielding materials has been described. Specimens with windows, rings, and holes were used to determine the effects of installation or in-service damage on corrosion of shielding materials. The effect of alternating current was

TABLE 5—*Rough indications of soil corrosivity versus resistivity.*

Resistivity, ohm-cm	Corrosivity, description
Below 500	Very corrosive
500 to 1000	Corrosive
1000 to 2000	Moderately corrosive
2000 to 10 000	Mildly corrosive
Above 10 000	Progressively less corrosive

TABLE 6—*A-C current in shields of test cables.*

Shield Type	Shield Current at Beginning of Test Cable, mA	Shield Current at End of Test Cable, mA
ASP	100	180
CASP	180	130
CACSP	130	205
Cu/ST/Cu	205	215
Al/ST/Al	215	200
CAl/ST/CAl	200	150

TABLE 7—*Rating code for the evaluation of shields in cable specimens.*

Rating	Performance	Degree of Corrosion
10	Excellent	Unaffected—no indication of corrosion
9	Excellent	Superficial rust or etching on surface
8	Very good	Uniform metal attack, rust, and/or slight localized pitting
7	Good	Appreciable pitting over the surface, but no perforations through metal shield; some minor delamination or dissipation of metallurgically or plastic-bonded metals leaving cathodic metal intact
6+	Good	Localized pitting: only one perforation in shield by pitting
6	Good	Localized pitting: two to five perforations in shield by pitting
5	Fair	Many localized pits causing perforation of shield <5% of shield dissipated by corrosion; extensive delamination of metallurgically bonded metals
4	Poor	Severe corrosion: pitting to perforation of shield; 5 to 10% of shield dissipated by corrosion; severe corrosion of anodic part of metallurgically bonded metals
3	Poor	Severe corrosion: pitting to perforation of shield; 10 to 25% of shield dissipated by corrosion
2	Very poor	Severe corrosion: more than 25% of shield dissipated by corrosion; shield still has electrical continuity along the cable
1	Very poor	Severe corrosion: shield is close to electrical discontinuity (ELD) due to perforation in shield and dissipation of metal by corrosion
0	Very poor	Severe corrosion: shield is electrically discontinuous (ELD) due to dissipation of metal by corrosion

TABLE 8—*Performance of cable-shielding materials after 0.5-year exposure (Conway, SC).*

Sample	Static	Dynamic	Window	Ring	Holes 1/4	1/8	1/16	1/32	Under Jacket
ASP	Steel		6	5	8[1]	4[1]	7[1]	10[1]	8
	Al		(10)	(10)	(10)	(10)	(10)	(10)	(10)
		Steel	5	2	7	9	9	9	8
		Al	(10)	(10)	(10)	(10)	(10)	(10)	(10)
CASP	Steel		6	5	10	10	10	10	9
	Al		(10)	(10)	(10)	(10)	(10)	(10)	(10)
		Steel	5[1]	1	5[1]	5[1]	0[1]	10	9
		Al	(10)[2]	(10)	(10)	(10)	(10)	(10)	(10)
CACSP	Steel		9[1]	9[1]	6+[1]	6+[1]	0[1]	9	9
	Al		(10)	(10)	(10)	(10)	(10)	(10)	(10)
		Steel	6[1,3]	6[1,3]	8[1]	9[1]	0[1]	10	9
		Al	(10)	(10)	(10)	(10)	(10)	(10)	(10)
Cu/ST/Cu	Cu		9[1]	9[1]	9[1]	9[1]	10[1]	10[1]	9[1]
	ST		9	...	...	...	...	...	...
	Cu		(9)[1]	(9)[1]	(10)	(10)	(10)	(10)	(10)
		Cu	9[1]	9[1]	9[1]	9[1]	10[1]	10[1]	9[1]
		ST	9	...	...	...	...	...	...
		Cu	(9)	(9)	(10)	(10)	(10)	(10)	(10)
Al/ST/Al	Al		10	10	10	10	10	10	10
	ST		8	...	...	...	...	...	...
	Al		(10)	(10)	(10)	(10)	(10)	(10)	(10)
		Al	10	10	10	8[1]	10	10	9
		ST	8	...	...	...	...	...	...
		Al	(10)	(10)	(10)	(10)	(10)	(10)	(10)
CAl/ST/CAl	Al		10	10	10	10	10	10	10
	ST		10	...	...	...	...	...	...
	Al		(10)	(10)	(10)	(10)	(10)	(10)	(10)
		Al	10	10	10	10	10	10	9
		ST	9	...	...	...	...	...	...
		Al	(10)	(10)	(10)	(10)	(10)	(10)	(10)

NOTE: For explanation of footnotes, see Table 13.

TABLE 9—*Performance of cable-shielding materials after 1.0-years exposure (Conway, SC).*

Sample	Static	Dynamic	Window	Ring	Holes 1/4	1/8	1/16	1/32	Under Jacket
ASP	Steel		3	3	6[1]	...	...	...	8
	Al		(10)	(10)	(10)	(10)	(10)	(10)	(10)
		Steel	2	1	7	10	10	10	7
		Al	(10)	(10)	(10)	(10)	(10)	(10)	(10)
CASP	Steel		5[1]	5[1]	5[1]	6+	10	10	8
	Al		(10)	(10)	(10)	(10)	(10)	(10)	(10)
		Steel	3[1]	2[1]	0[1]	0[1]	5	5	8
		Al	(10)	(10)	(10)	(10)	...	...	(10)
CACSP	Steel		8[1]	8[1]	7	6[1]	7[1]	7[1]	10,7
	Al		(10)	(10)	(10)	(10)	(10)	(10)	(10)
		Steel	5[1,6]	6[1]	8[1]	7	5[1]	0[1]	10,7
		Al	(10)	(10)	(10)	(10)	(10)	(10)	(10)
Cu/Steel/ Cu	Cu		9	9	7	10	10	10	10
	ST		8	...	...	...	...	...	...
	Cu		(10)	(10)	7	(10)	(10)	(10)	(9)
		Cu	9	9	7	9	9	10	9
		ST	10	...	...	...	...	...	...
		Cu	(10)	(10)	(10)	(10)	(10)	(10)	(10)
Al/ST/Al	Al		10	10	7	10	10	10	10
	ST		8	...	...	...	...	...	...
	Al		(10)	(10)	7	(10)	(10)	(10)	(10)
		Al	10	10	7	10	10	10	10
		ST	7	...	...	...	...	...	...
		Al	(10)	(10)	7	(10)	(10)	(10)	(10)
CA1/ST/CAl	Al		10	10	7	10	10	10	10
	ST		10	...	...	...	...	...	...
	Al		(10)	(10)	7	(10)	(10)	(10)	(10)
		Al	10	10	7	10	10	10	10
		ST	10	...	...	...	...	...	...
		Al	(10)	(10)	7	(10)	(10)	(10)	(10)

NOTE: For explanation of footnotes, see Table 13.

TABLE 10—*Performance of cable-shielding materials after 2.5-years exposure (Conway, SC).*

Sample	Static	Dynamic	Window	Ring	Holes 1/4	1/8	1/16	1/32	Under Jacket
ASP	Steel		1	0	4	7	7	8	8
	Al		(10)	(10)	(10)	(10)	(10)	(10)	(10)
		Steel	4	0	7	7	7	7	9
		Al	(10)	(10)	(10)	(10)	(10)	(10)	(10)
CASP	Steel		5	4	5	4	10	10	9
	Al		(10)	(10)	(10)	(10)	(10)	(10)	(10)
		Steel	2	0	0	2	3	3	8
		Al	(10)	(10)	(10)	(10)	(10)	(10)	(10)
CACSP	Steel		6	8[1]	8[1]	8[1]	0[1]	10	10
	Al		(10)	(10)	(10)	(10)	(10)	(10)	(10)
		Steel	5[8]	5	4[1]	7	0[5,8]	0[5,8]	10
		Al	(10)	(10)	(10)	(10)	(0)[5]	(0)[5]	(10)
Cu/ST/Cu	Cu		8[9]	8[9]	8[2]	8[2]	8[2]	8	9
	ST		3	3	...	...	...	...	...
	Cu		(8)	(9)	(10)	(10)	(10)	(10)	(9)
		Cu	9	9	9[1]	9	9	9[1]	9
		ST	9	7	...	...	...	...	...
		Cu	(9)	(9)	(10)	(10)	(10)	(10)	(9)

NOTE: For explanation of footnotes, see Table 13.

TABLE 11—*Performance of cable-shielding materials after 4.0-years exposure (Conway, SC).*

Sample	Static	Dynamic	Window	Ring	Holes, in. 1/4	1/8	1/16	1/32	Under Jacket
ASP	Steel		0	0	...	...	...	...	5
	Al		(10)	(6+)	(10)	(10)	(10)	(10)	(10)
		Steel	1	0	0	0	...	...	7
		Al	(10)	(10)	(10)	(10)	(10)	(10)	(10)
CASP	Steel		1	1	0	6+	...	...	6
	Al		(10)	(10)	(10)	(10)	(10)	(10)	(10)
		Steel	0	0	0	0	0	...	5
		Al	(10)	(10)	(10)	(10)	(10)	(10)	(10)
CACSP	Steel		10	10	10	10	10	10	10
	Al		(10)	(10)	(10)	(10)	(10)	(10)	(10)
		Steel	8[10]	6[10]	...	0	0	0	9[10],7[8]
		Al	(10)	(10)	(10)	(10)	(10)	(10)	(10)
Cu/ST/Cu	Cu		8[9]	8	9	9	9	9	9
	ST		7	7	...	...	...	9	9,7[8]
	Cu		(10)	(10)	(10)	(10)	(10)	(10)	(10)
		Cu	8	8	8[2]	8[2]	8[2]	8[2]	9
		ST	8	8	...	...	...	...	9
		Cu	(8)	(9)	(10)	(10)	(10)	(10)	(10)
Al/ST/Al	Al		6+	6	10	10	10	10	10
	ST		7[8]	7[8]	10	10	10	10	9,7[8]
	Al		(10)	(10)	(6+)	(10)	(10)	(10)	(10)
		Al	10	10	10	10	10	10	10
		ST	7[8]	7[8]	10	10	10	10	9,9[8]
		Al	(10)	(10)	(10)	(10)	(10)	(10)	(10)
CAl/ST/CAl	Al		10	10	10	10	10	10	(10)
	ST		10	10	10	10	10	10	10
	Al		(10)	(10)	(10)	(10)	(10)	(10)	(10)

TABLE 11—*Continued.*

Sample	Static	Dynamic	Window	Ring	Holes, in.				Under Jacket
					$1/4$	$1/8$	$1/16$	$1/32$	
		Al	6	5	10	10	10	0^2	10
		ST	6	5	10	10	10	0^2	9^8
		Al	(10)	(10)	(10)	(10)	(10)	(10)	(10)

NOTE: For explanation of numbers in columns and in footnotes, see Table 13.

TABLE 12—*Performance of cable-shielding materials after 6.0-years exposure (Conway, SC).*

Sample	Static	Dynamic	Window	Ring	Holes				Under Jacket
					$1/4$	$1/8$	$1/16$	$1/32$	
ASP	Steel		0	1	0	0	0	10	7
	Al		(6)	(5)	(10)	(10)	(10)	(10)	(5)
		Steel	1	1	3	10	10	10	9
		Al	(10)	(10)	(10)	(10)	(10)	(10)	(10)
	Steel		1	1	3	0	10	10	7
	Al		(5)	(5)	(3)	(10)	(0)	(10)	(5)
		Steel	1	1	0	7	10	10	9
		Al	(10)	(10)	(10)	(10)	(10)	(10)	(10)
	Steel		0	0	1	4	0	3	7
	Al		(9)	(3)	(10)	(10)	(10)	(10)	(8)
CASP	Steel		0	0	0	0	0	0	5
	Al		(10)	(10)	(10)	(10)	(10)	(10)	(10)
	Steel		0	0	0	0	0	0	5
	Al		(10)	(10)	(10)	(10)	(0)	(10)	(10)
		Steel	0	0	0	0	0	0	5
		Al	(10)	(10)	(10)	(10)	(10)	(10)	(10)
CACSP	Steel*		10	10	0	0	0	0	10
	Al*		(10)	(10)	(10)	(10)	(10)	(10)	(10)
		Steel	4^{10}	1	3	10	10	10	5^{10}
		Al	(10)	(10)	(10)	(10)	(10)	(10)	(10)
	Steel		6^{10}	6	1	1	1	1	10
	Al		(10)	(10)	(10)	(10)	(10)	(10)	(10)
		Steel	5	7^8	8^{10}	9	10	10	10
		Al	(10)	(10)	(10)	(10)	(10)	(10)	(10)
Cu/ST/Cu	Cu		9	9	9	9	9	9	10
	ST		5^8	7^8	...	...	...	...	...
	Cu		(10)	(10)	(10)	(10)	(10)	(10)	(10)
		Cu	9	9	9	9	9	9	10
		ST	9	...	...	...	...	...	...
		Cu	(9)	(9)	(10)	(10)	(10)	(10)	(10)
	Cu		9	9	9	9	9	9	10
	ST		5^8	5^8	...	5	...	...	...
	Cu		(10)	(10)	(10)	(10)	(10)	(10)	(10)
		Cu	9	9	9	9	9	9	10
		ST	10	10	...	...	...	...	...
		Cu	(9)	(9)	(10)	(10)	(10)	(10)	(10)
Al/ST/Al	Al		6	6	8^2	10	10	10	6+
	ST		10	6	...	...	...	...	...
	Al		(8)	(6)	(10)	(10)	(10)	(10)	(10)

TABLE 12—*Continued.*

Sample	Static	Dynamic	Window	Ring	Holes 1/4	1/8	1/16	1/32	Under Jacket
		Al	3	4	9	9	9	9	7
		ST	5	6+	. . .	. . .	. . .	. . .	. . .
		Al	(5)	(5)	(10)	(10)	(10)	(10)	(7)
	Al		8	8	10	10	10	10	10
	ST		10	10	. . .	. . .	. . .	. . .	. . .
	Al		(9)	(10)	(10)	(10)	(10)	(10)	(10)
		Al	6	6	4	6+	8^2	10	7
		ST	6	6	4	4	8	. . .	. . .
		Al	(6)	(6)	(6+)	(10)	(10)	(10)	(7)
CAI/ST/CAI		Al	6	6	10	10	10	10	10
		ST	6	6	. . .	. . .	. . .	. . .	. . .
		Al	$(6+)^8$	$(6+)^8$	(10)	(10)	(10)	(10)	(10)
	Al		6^{10}	10^2	10	10	10	10	10
	ST		8	10^2	. . .	. . .	. . .	. . .	. . .
	Al		(10)	(10^2)	(10)	(10)	(10)	(10)	(10)
		Al	5	5	10	10	10	10	10
		ST	6+	6+	. . .	. . .	. . .	. . .	. . .
		Al	(7)	(9)	(10)	(10)	(10)	(10)	(10)

NOTE: For explanation of footnotes, see Table 13.
*Unearthed by CATV Co. and reburied.

TABLE 13—*Explanatory notes.*

1. Corrosion at/or along line where shield was touched by punch or cutting tool.
2. Mechanical damage.
3. Coating torn.
4. NBSIR 81-2243, 4/81.
5. Sample area missing probably due to preparation of holes with punch.
6. One perforation not initiated by mechanical damage.
7. Removed for metallurgical analysis by Texas Instruments prior to exam by panel.
8. Extensive corrosion from the edge.
9. Mechanical deformation (dimple) caused by internal corrosion products.
10. Filliform corrosion observed:
    - Inner jacket evaluation indicated by ( ).
    - Blanks indicate inability to evaluate.
    - Where dual ratings are given, the second rating is for edge effects.
    - Evaluations are based on exposed area only.

determined by comparing results from isolated specimens to those of specimens electrically connected to the cable in service.

Chemical characterization of the soil in Conway, SC was accomplished by preparing soil water extracts. Cation analysis was accomplished with atomic absorption spectroscopy, while anion analysis required the ion specific techniques listed in Table 2. The pH, resistivity, and chemical composition from the test site indicated that it was similar in corrosivity to Hagerstown loam. The results from these tests as judged by a panel of industry representatives have been listed.

Acknowledgments

The authors wish to express their appreciation to REA, Contel Corp., The Dow Chemical Co., and Texas Instruments Inc. for their participation in the test program and Horry Telephone Cooperative for installing and retrieving specimens.

References

[1] Fisher, E. L., Robinson, E. A., and Bishop, W. F., "Lightning Shielding of Plastic Telephone Cable," *Proceedings,* 17th International Wire & Cable Symposium, Atlantic City, NJ; 5 Dec. 1968.
[2] Connolly, R. A. and Landstrom, R. E., *Materials Research and Standards,* Vol. 9, 1969, p. 13.
[3] Connolly, R. A. and Cogelia, N., "The Gopher and Buried Cable," *Bell Laboratories Record,* April 1970.
[4] Backlund, J. B., "Are Gophers Gaining in Guerrilla War," *Telephone Engineer and Management,* 1969, p. 83.
[5] Masciarelli, A. and Savolainen, U. U., "Clad Metal Longitudinally Corrugated Shields Applied to Control Cables," *Proceedings,* Conference on Electric Wire and Cable Technology, New York, NY, 1972.
[6] Raman, R. and Gerdes, R. J., "One Telco's View of Cable Shielding," *Telephony,* October 1985, p. 40.
[7] Baboian, R. and Haynes, G. S., "A Comparative Study of the Corrosion Resistance of Cable Shielding Materials," *Materials Performance,* National Association of Corrosion Engineers, Vol. 18, No. 2, February 1979, pp. 45–56.
[8] Baboian, R., Hartley, S. R., and Hyman, E. D., "High Strength Corrosion Resistant Clad Metal Shielding for Telephone Wire and Cable," *Proceedings,* 23rd International Wire & Cable Symposium, Atlantic City, NJ, 1974.
[9] Schwank, G. D. and Bow, K. E., "Galvanic Corrosion Studies of Aluminum and Steel Shielding Materials For Armored Telephone Cables," *Materials Performance,* Vol. 17, No. 9, September 1978.
[10] Gerhold, W. F. and Fink, J. L., "Corrosion Evaluation of Underground Telephone Cable Shielding Materials," NBSIR81-2243, prepared for the Rural Electrification Administration; National Bureau Of Standards, Washington, DC, April 1981.
[11] Gerhold, W. F., Escalante, E., and Fink, J. L., "Corrosion Evaluation of Underground Telephone Cable Shielding Materials," NBSIR 82-2509, prepared for the Rural Electrification Administration; National Bureau of Standards, Washington, DC, June 1982.
[12] Fink, J. L. and Escalante, E., "Corrosion Evaluation of Underground Telephone Cable Shielding Materials," NBSIR83-2702, sponsored by the Rural Electrification Administration; National Bureau of Standards, Washington, DC, May 1983.
[13] Denison, I. A., *Journal of Research of the National Bureau of Standards,* Vol. 17, September 1936.
[14] Franson, M. A. H., Ed., *Standard Methods for the Analysis of Water and Wastewater,* 15th ed., American Public Health Association, 1981.

Summary

The keynote speaker was John H. Fitzgerald, III, Vice President of PSG Corrosion Engineering. The title of his paper was "The Future as a Reflection of the Past." Scientists tell us that scientific knowledge doubles every ten years. Taking as his theme the fact that continued growth in scientific knowledge is based on the foundation laid by others, Fitzgerald traced the growth of corrosion control from the early 20th century to the present. He noted that around the turn of the century nearly all underground corrosion was attributed to stray current "electrolysis" from street railways and subways. In 1910, the National Bureau of Standards began a study of this "electrolysis" and by 1920 concluded that soil corrosion was equally as serious as stray current corrosion. So, in 1922 the study was expanded to cover soil corrosion. The parameters responsible for soil corrosion were evaluated through long-term burial tests, and the report was published in 1945.

Fitzgerald went on to discuss engineers and scientists who studied the effects of different soil parameters such as resistivity, acidity, bacterial action, and moisture. He outlined the contribution each researcher made to the understanding of underground corrosion and showed how their work became the building blocks of today's instrumentation, procedures, and technology.

"Soil corrosion is too complex to permit correlation with any one parameter," says the 1945 NBS report. He said that we know today how true this is, and through the use of statistics and other methods of analysis we attempt to establish the effects of a given soil on underground facilities. "But let us remember," he said, that our success today has been made possible by those who have contributed to it over the last 75 years."

Looking to the future, he pointed out the need for further understanding of the interaction of various soil parameters to enable the corrosion engineer to make more accurate predictions of their effect on pipelines, tanks, and the like. "Let us study the work done in the past, reflect on it, and build on it as we go forward into the future," he said.

David Palmer, president of Corrosion Control Engineering, Ltd., presented the second paper, entitled "Environmental Characteristics Controlling the Soil Corrosion of Ferrous Piping." He examined six characteristics of soils controlling the external corrosion of ferrous piping materials, with particular reference to the AWWA rating formula. Under the heading Material Performance, he reported that cast iron (pit-cast and centrifugally cast) has commonly given service life in the 100-year range. He said that the interpretation of cast iron failure data is difficult because most failures are described as "breaks," whether due to purely mechanical effects or partially due to the weakening effect of corrosion.

"Usually, the only leaks attributed to corrosion are those where there is an obvious blowout of the graphitized part of the pipe wall without an accompanying mechanical failure," he said.

"In the 1960s," he added, "ductile iron was widely used as a replacement for cast iron based on the understanding that its corrosion resistance property was equal or superior to gray cast iron. In the 1950s, steel-coated pipe protected by cathodic protection was introduced for its improved pressure rating and relative economy of installation."

In the section entitled Corrosion Morphology—Ferrous Materials, he mentioned that the corrosion of mild steel produces no particularly significant behavior other than the usual lowering of corrosion rate with time as the corrosion products introduce additional resistance in the corrosion cell electrical circuit. "On the other hand," he added, "the relationship of the cathodic graphite to the anodic iron in cast and ductile iron pipe has long been of interest, and the size and shape of the graphite particles in relation to corrosion resistance has been examined without firm conclusions reached. Metallurgical tests tend to confirm that the corrosion of cast iron is nucleated by graphite iron galvanic cells and suggest that the graphite/corrosion product deposit's pressure-retaining ability is influenced by the characteristics of the matrix established by the graphite flakes."

A review conducted by Canada's National Research Council concluded that the corrosion rate of all ferrous materials by soils is essentially equal.

In the section entitled Mitigative Action, he stated that to date it consisted mainly of logging clamped leaks and installing sacrificial cathodic protection anodes at each leak.

As for soil characteristics, he reviewed several parameters with respect to their reliability and relevance as corrosion indicators. "Resistivity," he said, "is a function of soil moisture and concentration of current-carrying soluble ions." He stated that the overwhelming majority of field studies show resistivity to be the controlling parameter except for areas with severe micro-biological activity.

In the section entitled pH, he mentioned that this criteria may be useful only in identifying unusual soil conditions.

In the section entitled Redox Potential, he said, "The redox potential parameter attempts to distinguish between aerobic and anaerobic soils," "Kuhlman and others have attempted with-out success to correlate redox potential with corrosion rate."

In the section entitled "Sulfides," he stated that sulfate levels are of significance where con-crete structures are considered.

In the Discussion section he explained the AWWA formula point system. He suggested to limit the parameters to be considered to two, leading to a requirement for protection when a resistivity is less than 1400 ohm-cm.

He concluded that resistivity mapping combined with pipe type/age plotting appears to be the most reliable approach to planning mitigative programs.

Paul A. Burda presented a paper entitled "Differential Aeration Effect on Corrosion of Cop-per Concentric Neutral Wires in the Soil." Field data showed that extensive localized corrosion cells on concentric neutral copper wires were associated with many failures. Differential aera-tion is considered by some investigators to be the most probable mechanism for this corrosion deterioration. Burda identified other factors affecting this corrosion phenomena: pH at the metal interfaces, the environment, the ohmic resistance, and the anodic reaction.

He reported that the results of laboratory experiments showed that the differential aeration effect increased the corrosion of copper in soil by 20 times. It was also found that chlorides in soils doubled the rate of copper corrosion. "When pH is between 6 and 8," he stated, "the differential aeration mechanism can control the corrosion of copper." The maximum rate of corrosion of concentric neutral copper wires in the soil was found when the anode to cathode ratio was 1:1. He also stated that anodic polarization, which may occur at high corrosion rates, can be neglected in ordinary cases.

Goran Camitz and Tor-Gunnar Vinka presented a paper entitled "Corrosion of Steel and Metal-Coated Steel in Swedish Soils—Effects of Soils Parameters." The paper presented the results of a long-term study being conducted by the Swedish Corrosion Institute. Carbon steel, zinc-coated steel, and aluminum-zinc alloy (trade name Galvalume) specimens in flat bar and plate forms were tested. Since groundwater was only 1 m from the surface, two specimens were used, one placed at about 0.7 m depth, the second at about 1.7 m. The soils tested were clay, muddy clay, silty clay, peat, and sand.

Detailed site locations were reported with several soil parameters measured for each site. Also, the detailed weight loss method was described; the results were presented in graph and table forms.

The study concluded the following:

1. The corrosion rate of carbon steel is higher above the groundwater table; as for zinc-coated steel and aluminum zinc alloy, no obvious effect was observed.

2. There is a higher corrosion rate of carbon steel panels placed in homogeneous sand when compared with native soil. Zinc-coated panels were lower in corrosion rate in sand. Aluminum zinc alloy panels reported a similar tendency as zinc, with a less drastic rate of change as in the case of zinc.

3. The pitting rate of carbon steel panels is considerably high in sand above the groundwater table.

4. In general, the corrosion rate is high in soils with low pH for all tested specimens.

5. Muddy clay and peat have the highest corrosivity rate for all three materials, and sand has the comparatively lowest rate.

6. The corrosion rate is the same in both sample shapes, although the plates have 15 times larger exposed area to the flat bars.

7. The corrosion rates of carbon steel are relatively constant with time.

Robert C. Rabeler presented a paper entitled "Soil Corrosion Evaluation for Screw Anchors." The paper described an evaluation performed over a seven-year period to evaluate corrosion of guy anchors for a transmission power line. Galvanized screw anchors were used. Instantaneous corrosion rates were evaluated using polarization testing techniques, and actual thickness and weight loss measurements were performed to verify the results.

He pointed out that both the linear polarization and polarization break techniques can be used to calculate corrosion current. Faraday's law must then be used to convert from corrosion current to corrosion rate.

The paper described in detail and step-by-step each decision and the reason for it. While the test anchors were in the ground, polarization tests were performed; both linear polarization and polarization break techniques were used. After the removal of the anchors, weight loss and corrosion rates were calculated; pit depth was also measured.

In the Discussion section, Rabeler reported that the galvanized steel bolts indicated weight losses ranging from 0.1 to 0.2% of the original weight. He also indicated loss in total zinc thickness of approximately 0.5 to 2.0 mils after 4.3 years. This translates to an annual average corrosion rate from 0.06 to 0.23 mils per year.

Polarization test measurements accurately predicted the actual measured corrosion rates. The polarization break technique appeared to most accurately reflect actual weight loss measurements. The study did not show a clear trend in corrosion rate with depth. In conclusion, Rabeler confirmed that polarization tests can be performed to accurately predict corrosion rates of buried metallic structures.

Edward Escalante presented a paper entitled "Concepts of Underground Corrosion." He defined underground corrosion as the deterioration of metals, or other materials, brought about by the chemical, mechanical, and biological action of the soil environment.

In the section Basic Concepts, Escalante stated that underground corrosion is electrochemical in character; thus, the corrosion process can be examined by electrical means. He said the process is very similar to the electrochemical action that takes place in an ordinary dry cell. In his explanation he mentioned that the anode goes into solution in the electrolyte; this dissolution is referred to as an oxidation reaction. On the other hand, reduction reactions occur at the cathode, leading reduced ions such as hydrogen to adhere to the cathode surface and stop further reaction. The driving force for any galvanic cell is the potential difference between the anode and the cathode. The difference in potential developed between two metals and their relative chemical performance can be judged by examining a galvanic series. Differences in grain orientation can cause some grains to act as anodes while others act as cathodes with excellent electrical continuity existing in the bulk material. Inhomogeneities in the electrolyte can also cause potential difference on a metal surface.

In the section Corrosion in Soil, Escalante gave the Department of Agriculture definition of soil as the loose surface material on the earth consisting of disintegrated rock with an admixture of organic material on which plants grow. The corrosion behavior of structural steel in soil can be divided into two categories, corrosion in disturbed soil and corrosion in undisturbed soil.

In the section Corrosion in Disturbed Soil, Escalante identified some of the factors the National Bureau of Standards (NBS) has evaluated over a period of years from the seven underground corrosion test sites in the United States:

1. *Soil texture,* which is determined by the proportions of sand, silt, and clay that make up a soil. It has an important influence on the diffusivity of soluble salts and gases.

2. *Internal drainage* is the property that describes the water retention of a soil.

3. *Soil resistivity* is a measure of how easily a soil will allow an electric current to flow through it.

4. *Temperature of soil;* it does not have as large an effect on underground corrosion as one might expect.

5. *Soil pH* is the acidity or alkalinity of the soil media. It has little effect on corrosion of steel.

6. *Redox potential or oxidation*—reduction potential is the potential of a platinum electrode versus a reference half-cell converted to the hydrogen scale. It is an indication of the proportions of oxidized and reduced species in a specific soil.

In the Summary section, Escalante concluded that corrosion in disturbed soil is a function of the soil environment, but soil pH and redox potential are poor indicators of a corrosive soil. A soil with a resistivity below 500 ohm-cm is corrosive. Above 2000 ohm-cm, the relation of soil resistivity to soil corrosivity is less reliable.

He noted that the corrosion of steel piles in undisturbed soil is independent of the soil environment. Even with low soil resistivities, the corrosion observed is very low. Coating the cathodic area of the pile in the disturbed soil zone above the water line or in the concrete cap will further reduce corrosion effects.

Richard A. Corbett and Charles F. Jenkins presented a paper entitled "Soil Characteristics as Criteria for Cathodic Protection of a Nuclear Fuel Production Facility." They used leak frequency curves from other nearby plant sites, extensive soil resistivity surveys, and geochemical analyses to evaluate the onsite soil characteristics for corrosion susceptibility.

The paper recounted the steps taken to investigate soil corrosivity and to determine the extent of the necessary corrosion control measures.

The Defense Waste Processing Facility is designated to receive radioactive wastes from the Savannah River Plant nuclear fuel production in a liquid slurry form and encapsulate it into a permanent solid glass form. The wastes from the chemical separations process and tank form storage areas will be transferred through underground piping systems up to five miles. Because of the radioactive nature of the slurry, special care utilizing conservative design and installation approaches are applied throughout. Public safety demands assurance that no failures occur during the reasonable design life of the entire system.

The paper stated that the soil represents the last controllable means of protection against contamination of the water table and nearest aquifer. Low permeability, impervious clay provides the slowdown of percolation, which is desired. This is due to the characteristics of clay, namely absorption of water, swell, and ion exchange. There is a negative effect in the tendency of wet clay to hold moisture in the vicinity of buried lines. If the soil is high in soluble salts or if it has high total acidity and is alternately wet and dry, it may be especially corrosive.

The result of the leak frequency curve for the site under investigation is classical in nature, he noted, and follows general experience with underground corrosion. This means that once leaks start, an increase in their rate of development can be anticipated.

Another interesting factor mentioned was that disturbance of soil, disturbance of compaction, and use of heavy equipment all contribute to failures in cast iron piping and can sometimes be related to later corrosion occurrences in an area.

The paper stated that the most commonly agreed-upon criteria to rank the degree of corrosivity among soils are resistivity and total acidity. Large variations in soil resistance provide for a possibility of galvanic couples.

Long line corrosion usually occurs when the pipe traverses soils of different composition (for example, one section of the pipe becomes anodic with respect to another).

The paper points out that cathodic protection was recommended for the Defense Waste Processing Facility project based on a conservative approach, including:

1. Heterogeneous soil resistivities that could lead to galvanic corrosion.
2. Soil chemistry leading to a corrosive tendency.
3. A leak frequency history in adjacent areas.

The cathodic protection system was an impressed current of closely distributed anodes to limit the amount of current discharge per anode to reduce voltage gradients around the anodes, leading to a minimum of detrimental effects of stray currents occurring on electrically discontinuous structures.

James B. Bushman and Thomas E. Mehalick presented a paper entitled "Statistical Analysis of Soil Characteristics to Predict Mean Time to Corrosion Failure of Underground Metallic Structures."

The paper started by identifying Ohm's law as the law that corrosion rate of buried or submerged metallic structures follows. Corrosion current is directly proportional to the voltage of the corrosion cell and inversely proportional to the resistance of the corrosion cell.

The authors mentioned that for a number of years corrosion engineers have been using structure-to-electrolyte and other electrical potential measurement techniques to analyze corrosion patterns on underground pipelines. These did not help determine the rate of time to future failure.

The paper enumerated research work using statistical analysis starting with Gordon Scott, who determined that soil resistivity was "normally" distributed if the logarithm of the resistivity was used in the analysis. This was followed by the Husock and Wagner evaluation, the probability of corrosion leaks versus the logarithm of soil resistivity. This, in turn, was followed by Warren Rogers, who developed a computer model which could predict the mean time to corrosion failure (MTCF) for each site. The paper identified some eleven soil characteristics that impact the corrosion rate of buried metallic structures and eight structure factors.

The authors used multivariate and nonlinear regression analysis to develop a mathematical model for predicting the MTCF, which is the average age at which each location will leak due to corrosion. When the model was tested in a pipeline study, it resulted in a coefficient of determination value in excess of 0.95, which is considered to be extremely high. The authors also showed by their newly developed model the inability of using any single soil characteristic to predict MTCF.

K. P. Fisher and O. R. Bryhn presented a paper entitled "Corrosion and Corrosion Evaluation of Superficial Sediments on the Norwegian Continental Shelf."

The purpose of the paper was to report the methodology used in evaluating the corrosivity of the Norwegian soil below sea water to a depth of 500 m. They said that geotechnical properties of the soil can be considered reasonable data to determine the corrosivity of the soil. For more accurate information and corrosion rates, detailed electrochemical studies are necessary. The authors stated that "the marine sediments are mainly anaerobic and the activity of the sulphate reducing bacteria has been considered to be the main cause for free corrosion." They followed King's scheme for corrosion prediction, which is based on sediment type, organic content, water depth, sea water content of nitrogen and phosphorous, and temperature. They found that the main part of the Norwegian sector of the North Sea is low in corrosivity with the exception of the southern coastal area of Norway, where sulfate-reducing bacteria can be expected.

The authors described in detail the method of sampling soil, the differences which can be encountered between laboratory versus in situ for exposure of metal samples, and several other factors important to collect reliable data. The factors they considered in their study were:

1. Corrosion caused by sulfate-reducing bacteria.
2. The influence of precipitated ferrous sulfides on the cathodic and anodic reactions.
3. Formulation of protective or nonprotective ferrous sulfides on the cathode.
4. Formation of galvanic cells due to the presence of ferrous sulfides.

The reported results were as follows:

1. No simple relationship between resistivity and the corrosion rate was found.
2. Steel surfaces prepared by grinding showed lower corrosion rates when compared with gritblasted surfaces.
3. The corrosion rate obtained by the galvanostatic polarization method was found to be two to three times higher than the one obtained by weight loss.
4. The laboratory evaluation produces much lower corrosion rates when compared to the in situ tests in cases of high activity of sulfides and/or hydrogen sulfide (H_2S).
5. The average value of current demand for cathodic protection in a marine sediment was found to be 17 mA/m^2.
6. A general trend was found of decreasing activity of sulfate-reducing bacteria with increasing water depth and distance from land.
7. The results of the in situ corrosion exposures performed have generally given very low corrosion rates.
8. The corrosion rate of steel in the Norwegian Continental Shelf can be expected to be very low in most areas. In some areas, it could be very corrosive due to organic material, which can be distinguished by a strong smell of amines.

Thomas V. Edgar presented a paper entitled "In-Service Corrosion of Galvanized Culvert Pipe." The paper showed clearly that minimum resistivity and soil pH, the two most commonly used soil parameters for culvert pipe selection, may be inappropriate values to use in practice.

The author stated that "the most common reason for a culvert pipe to fail is due to a gradual weakening caused by corrosion."

Minimum resistivity of the soil and the pH of the soil and water are the two parameters used by many states to estimate the years to perforation of 16-gauge steel culvert. These and other parameters were studied both in situ and in the laboratory for twelve 35 to 40-year-old culvert pipes from four highway reconstruction sites in Wyoming to determine corrosion protection criteria.

Under general observations, Edgar reported that the corrosion was usually found in local areas on the pipe, indicating that small corrosion cells damaged the most pipe. He added that the most corroded area was found in the center of the culvert and the most anodic area in an oxygen concentration cell under a roadway. He also observed that the band joining two sections of pipe in the centerline of the road experienced the worst corrosion. The invert of the pipe, considered anodic to the crown, suffered the worst corrosion.

The author developed a mathematical relationship to predict weight loss using the field resistivity. He confirmed previous findings that pH of the soil has little or no effect on the corrosion when it is between 6.0 and 9.0. He found that the data indicate a reasonable correlation between the percent of soluble salts and minimum resistivity.

Edgar concluded by stating that "the most obvious defect in practice was the poor backfill material used." He recommended the use of a clean, coarse, cohesionless backfill material.

Gardner Haynes, Gregory Hessler, Reiner Gerdes, Kenneth Bow, and Robert Baboian presented a paper entitled "A Method for Corrosion Testing of Cable-Shielding Materials in Soils." The paper identified mechanical strength, electrical conductivity, and corrosion resistance as important criteria for cable-shielding materials. It also identified the following for proper function: provide mechanical protection to the cable core during and after installation; prevent ingress of moisture into the cable core; and provide electrical conductivity for the life of the cable.

A previous study was conducted by NBS-REA; however, the study did not evaluate the comparative behavior of idle cable versus cable-carrying alternating current. The present study determined the corrosion behavior and the associated effects of alternating current that is present in service or in commonly used shielding materials.

The paper said that continuous 500-ft lengths of cable with different shielding materials were prepared by shielding manufacturers with damage sites at 30-ft intervals, as well as individual 2½-ft lengths of cable with damage sites. The damage site pattern was intended to simulate possible construction, lightning, or rodent damage to the cable's outer jacket. The paper described in detail the installation of cables and marking for ease of retrieval. The test program called for retrieval of a control cable (static) and a section of test cable (dynamic) of each type of shielding at intervals that varied from 0.5 to 6 years. The soil was tested for pH, electrical resistivity, Ca, Mg, Na, CO_3, HCO_3, SO_4, Cl, and NO_3. The a-c current in shields was measured at the beginning and at the end of determined test time.

The specimens were rated by a panel using a rating system with a scale of 0 to 10 with 10 being no indication of corrosion and 0 being electrical discontinuity due to corrosion. Results are reported, but since the ratings were the consensus opinion of a panel, explanatory notes were required.

The intent of the paper was to record these results in the literature without bias, and, therefore, no discussion or interpretation of these results is included.

Victor Chaker

The Port Authority of New York and New Jersey, Jersey City, NJ 07310-1397; symposium cochairman and editor

Author Index

Subject Index

A

A-c current, cable-shielding materials corrosion, 149, 154
Acidity (*See also* pH)
soil corrosion, 2
Acid rain, soil pH, 11
Aerated half cell, differential aeration experiment, 19–20
Aluzink-coated steel
corrosion rates, 43, 47
groundwater levels, 47–48
sandfill, 49–50
Swedish soils study, 40–42
Anisotropy, metal surface, 83–84
Anode/cathode system
underground corrosion concepts, 82–84
undisturbed soil corrosion, 91, 94
Anodic/cathodic polarization
chloride-sulfate presence, 21–23
differential aeration, 19–20
Anodic corrosion rate, differential aeration, 25–26, 30–31
Asbestos cement, material performance, 6
ASTM Recommended Practices, G 1–81: 20–21
ASTM Standards
A 153: 54
C 105–72: 7, 12
D 1141–52: 32–35
D 2487: 85
G 57–84: 86
ASTM Test Methods
A 247–67: 7
G 51–77: 14, 86, 146
G 57–58: 11, 13, 56, 134, 146
Austenitic stainless steel, radioactive waste pipeline, 96
AWWA rating formula, 12, 16, 157
table, 7

B

Backfill
copper corrosion, 22, 24–25
culvert piping corrosion, 142
differential aeration experimental model, 18–19
radioactive waste pipeline, 96
Bacteriological corrosion, 3

Bag samples, culvert piping corrosion, 135
"Bell hole" excavation techniques, 112
Bituminous metal piping, 133
Brushing technique for culvert piping corrosion, 135–136

C

Cable-shielding materials
corrosion testing, 144–155
damage site patterns, 145–146
performance
0.5-year exposure, 150
1.0-years exposure, 151
2.5 years exposure, 152
4.0 years exposure, 152–153
6.0 years exposure, 153–154
rating, 150
Carbon steel
corrosion rate
as function of time, 49, 52–53
3 yrs. exposure, 42, 44
4 yrs. exposure, 42–43
pitting rate, 43, 45
radioactive waste pipeline, 96
Cast iron piping
corrosion morphology, 6–7
material performance, 5–6
mean time to corrosion failure (MTCF), 112–117
Cathodic passivation, differential aeration, illus., 21–22
Cathodic protection
DWPF design, 104–106
marine sediments, 128–129
Norwegian continental shelf, 119–120
soil characteristics, 95–106
soil survey, 103–104
Cathodic Tafel extrapolation, 127–128
Cation analysis, cable-shielding materials corrosion, 146–147, 154
Chemical impact on corrosion rate, 108–109
Chloride ion content
cable-shielding materials corrosion, 146–147
corrosion rates, table, 111
differential aeration, 21, 23, 25–27
ferrous piping corrosion, 16